THE
INTUITIVE ENGINEER
MEMOIR OF AN AEROSPACE · ENGINEER/ROCKET · SCIENTIST

JULIAN H. LINDER

DENVER, COLORADO

The Intuitive Engineer
Memoir of an aerospace-engineer/rocket -scientist
All Rights Reserved.
Copyright © 2015 Julian H. Linder
v2.0 r1.0

Outskirts Press, Inc.
http://www.outskirtspress.com

ISBN: 978-1-4787-6269-0

Library of Congress Control Number: 2015916132

Outskirts Press and the "OP" logo are trademarks belonging to Outskirts Press, Inc.

PRINTED IN THE UNITED STATES OF AMERICA

This book is dedicated to the engineers and scientists who shared their knowledge and expertise with me and to all of those who will follow in our footsteps

"I'm sorry, but this job *does* require you to be a
rocket scientist. . . ."

Contents

Introduction

THERE ARE NOT many books that have been written about engineers. As one who spent some 35 years working as an engineer, I believe that the lack of interest in the profession can be attributed to a large degree to it not being well understood and to a general ignorance regarding what an engineer actually does. For me, it was certainly exciting on many occasions and often extremely challenging. I am motivated to write this narrative by a desire to shed some light on the activities of the profession for those who might consider engineering as a career, as well as, to inform those who have an interest in the subject. I do this not as a recruitment effort but to convey a better sense of what is involved with the objective of dissuading those who are unsuited for the profession as much as to convince others that they might find intellectual fulfillment and career satisfaction in the field of engineering.

There are many forms of engineering such as electrical, civil, mechanical, etc. My career was in aerospace engineering and this is, therefore, the area with which I am most familiar and the one that I will describe and discuss. Companies that hire people to work as aerospace engineers look for college graduates with degrees in the sciences and favor those with educational backgrounds in mathematics, physics, electrical engineering and a few other specialties which have become prominent such as: aeronautical and aerospace engineering. I found it interesting that an educational background in one of the aforementioned disciplines does not prepare one to perform useful work. A period of training in the specific and peculiar interests of one's employer is nearly always required. The difference between education and training may not be readily apparent. The best analogy on the difference that I have heard is this: Parents of a teenager might be a bit apprehensive to find their child was taking a course in sex education at the school but would be much alarmed if their offspring were enrolled in a course in sex training.

Educational institutions could provide training as part of the curriculum; however, it would be difficult to provide training that would target the needs of a specific employer since such needs are diverse and varied. What might be effective and helpful to the student is a course in engineering workplace problem solving. This is a process that begins by first identifying a set of requirements and also a deliverable. The deliverable is highly important as it identifies what is to be produced as an end product and that which will also serve as evidence that the task has been completed. The activity then proceeds to the research phase, wherein all available information and data such as company documents, magazines, books and applicable documents in the public record are acquired and studied. At this point a list of possible approaches and courses of action should begin to come to mind. These are then evaluated and compared as to feasibility, effectiveness, cost, etc. A college course based on these guidelines would provide the student with a place to start on a first assignment in the workplace and avoid the embarrassment experienced by most new hires when one realizes that one doesn't know where to begin.

After a newly hired engineer has taken his or her first halting steps to accomplish an assignment that adds value to the project, a process of gathering knowledge and experience begins. This will take place over a number of years after which the engineer may be able to consider himself to be competent or even expert in a specific area and will then continue to expand his knowledge base and increase his ability to perform at higher levels. The mature engineer should be able to apply methods found in text books and the technical literature to solve current problems, as well as, use new approaches to problem solving from the latest research and advances in methodology. As the engineer solves a variety of problems, he will develop and accumulate a set of tools, which will be useful in solving future problems. The engineer also shares what he has learned with colleagues and contributes and adds to the company's expertise. At some point, the engineer should be able to quickly analyze the situation and develop solutions (if they exist) which will satisfy virtually any set of requirements that may come along. He is able to sense the solution and the path to be taken. At this point, he has then achieved the highest level of engineering expertise. He has become the "intuitive engineer."

Background and Education

THE DECISION TO pursue any given career is influenced by a number of things. First, of course, is one's ability and inclinations. Also, circumstances, the influences of various individuals such as parents, teachers, friends and acquaintances and the occurrences of random events will all have an effect.

In my case, I quickly caught on to the basics of mathematics, i.e., addition, subtraction, and multiplication. As early as the third grade, I was far enough ahead of my fellow students that the teacher enlisted me to drill the other students in reciting their addition tables. The ability to do simple arithmetic is an exercise in memorization and little more, although the early learner will probably develop some simple algorithms to help in memorizing sums, differences and products. One reason that arithmetic is found to be so tedious for most people is that it is not taught with any accompanying applications or uses. My fellow students and I plowed through this learning during our elementary school years and eventually came to the discipline of long division. I can still recall the endless hours spent grinding out the answers. It always seemed that a much smaller number of such problems would have sufficed.

By the eighth grade, we encountered the hated and dreaded "mixture problems." The learning potential of working such problems is actually quite high requiring the use of logic, visualization and working out the answer. Most students never got close to this. They were relieved to simply get past it without learning anything as long as their grades didn't take a big hit. I doubt if any of my teachers ever thought enough about "mixture problems" to make an inspired presentation. Perhaps they sensed that it was hopeless.

I was the middle child in a family of five children, three boys and two girls, well, until my youngest brother was born in the summer after I had finished high school. My parents had minimal involvement in our education. My father was a hopelessly addicted alcoholic who had only finished the eighth grade. My mother, God bless her, had graduated from high school and insisted that all her children would also do so. She actually had to fight a rather intense battle with our father so that my brother, the oldest, could attend the local high school. Once this battle had been won, the right to pursue a high school education was there for the rest of us. We were one of the poorest families in the small town where I grew up. Consequently, our parents had no influence in the community and chose to not be involved in any way in the running of the schools, which was probably a good thing for me and my siblings.

From an early age, I liked to build things using a hammer and nails. The produce crates that were discarded by the local supermarkets and stacked behind the building were a plentiful source of free wood. I dragged home many of these and built a playhouse and a go-kart, to name a few. I never got the Erector Set that I coveted (we were too poor), but I did get a cheap, scaled-down knock-off version one year at Christmas. It did have plates and girders and nuts and bolts, and I spent many hours building a variety of structures. I became interested in building model airplanes and built many rubber band powered and gas engine powered airplanes. I was also interested in real airplanes and subscribed to several aviation magazines. At the age of twelve, I had never flown in an airplane and was intensely interested in what it might be like. My nine year-old brother and I pooled our money and went down to the local airport where we hired a pilot to take us for a ride. We had eight dollars between us (actually, a lot of money back then), and apparently that was enough. They rolled out a four place Stinson Voyager and we had a 45 minute plane ride. I wondered what all the fuss was about, since the whole experience was rather boring. We seemed to hang motionless in the air.

Every high school freshman was required to take Algebra. The course had a reputation for being tough and most students were dreading the experience. The teacher was a rather eccentric but dear lady who insisted on opening wide all the windows in the classroom after each class period (apparently for health reasons), even in the dead of the Dakota winter. Algebra1 was for the most part a non-event, but the less capable students struggled. It was for them of little if any value, an experience to be soon forgotten. For those who were college bound, it was a necessary step, as it was a requirement for college admission. The curriculum for those continuing in the study of mathematics included geometry, advanced algebra and trigonometry. There was a separate, though informally defined, course of study for the college bound students that included the aforementioned mathematics regimen and also chemistry and physics. Latin (yes, Latin) was also recommended for the higher performing students. I began to follow the full mathematics curriculum. Unfortunately, the later courses were taught by a different teacher who delivered the course material in a sleep inducing monotone and never seemed to impart any enthusiasm for the subject material. Some might argue, possibly with some validity, that such is the nature of mathematics. I struggled through geometry and advanced algebra with this lady and could take no more. In my senior year, I opted for economics instead of trigonometry.

My chemistry teacher was exceptional and showed a lot of enthusiasm for the subject. In fact, when I took chemistry in college, it was like a review of what I had learned in his class. The physics teacher, on the other hand, was a dull, colorless individual who tolerated no levity of any kind in his classroom, apparently fearful that his authority would be challenged if he did so. My senior English teacher was the only teacher during my high school years that made any effort to encourage me. She suggested that I sign out to her room for last period study hall and then go to the public library (just across the street from the school) and read, which I did for the last semester of my senior year. Most of the books were old, but I found a few interesting titles. Among them was "Mien Kampf," a dull read, but I plowed my way through it. It seemed like a lot of nonsense, and I wondered what all the fuss had been about. There were, however, some interesting statements in it, and I decided to quote some of them in a presentation that I was to give in my International Relations class. I thought I would have a little fun with my classmates and present the quotes without attribution and see what reaction

I got before revealing where they came from. I couldn't have been more wrong and was totally unprepared for what happened. As fate would have it, there was a substitute teacher that day who just happened to be Jewish. I read the first quote and she exploded. "That's from 'Mein Kampf,' she shouted. I thought I was dead. Now realize that very few Jews, none that I knew of, lived in our town. My knowledge of contemporary Jewish culture was effectively nil. I did, however, have the benefit of having studied Jewish history as recorded in the Old Testament during my years in Lutheran Sunday school. I had a keen interest in Jewish history and a genuine sympathy for their suffering over the years. I had to work hard to convince the poor lady that I had great respect for the Jewish civilization and culture. After many statements to that effect, she finally relented and seemed satisfied that I was not promoting some neo-Nazi agenda.

After graduation, I had no plans to go to college in the fall. I had no money and no way to pay for college. I did, however, find a construction job that summer and saved a few hundred dollars. Near the end of the summer and as my bank account continued to grow, I began to think that college might be possible. I had enough money to get started and thought I might be able to work part time to pay for the rest of my expenses. I applied to the University of South Dakota in Vermillion, a small town about one hundred miles away. I was accepted and so, in early September, I loaded up my old Ford with my possessions and headed off to college.

I was surprised at how quickly my meager savings were eaten up by tuition, fees and books. Fortunately, the school offered to waive tuition because of my high score on the USD test that had been given to all high school seniors in the state the previous year (I had scored 11th out of almost 6000 seniors in the state). I registered as a journalism major because of my interest in reading and thinking that I would like to pursue a career in writing. This helped me get a part time job with the school News Bureau. The pay was below minimum wage and the place was run by a nasty fellow who thought it funny to make jokes about my financial circumstances and slender build. His favorite comment, "Have you had a good meal lately?" I found a few other jobs doing yard work, but these were insufficient to cover my expenses. I sold everything I owned: my shotgun, typewriter and car. I also petitioned my grandmother for help and the dear old lady sent me $500. I did make it through one semester but could clearly not go on.

I had to now choose a different life path but had no interest in joining the workforce and trying to live on a subsistence wage. It seemed like a couple of years in the US Army might be an interesting adventure, so I volunteered for the draft. Two months later I was on a bus heading for Camp Chafee, Arkansas.

This was in 1955. The Korean War was just over, and the training cadre were all veterans of that conflict. Some were also veterans of WWII. We trained pretty much for a rerun of those wars. The favored infantry weapon was the Garand M-1, a powerful and loud rifle. I put many rounds through mine and scored "sharpshooter" although I thought I should have scored "expert" because of my experience hunting pheasants during my high school years. Maybe because the guns were old and had gone though many training cycles, they were not as accurate as they might have been. In the first week of basic training, we were given a battery of tests covering many areas. I achieved a GA (general aptitude) score of 180. I had only a vague idea of what that meant, but if the mean was 100, that was pretty good. Anyway, this meant that I was qualified to apply for Officer Candidate School, which required filling out a number of forms with information I would have to write home for. The days were long and exhausting and I just couldn't find the energy to complete this task. It was, in retrospect, a good thing. One would have had to sign up for three years instead of the two I would be serving. Although it may have had a brief mention (I'll give the Army credit.), I was unaware at the time that an officer could be called back in to fight in any future wars. If I had known, that would have for sure been a deal breaker, and I would not have even considered OCS. Viet Nam, anyone? Basic training was tough but not harsh. Although time seemed to go by very slowly, the 8 weeks did eventually end. I was assigned to go to the Engineer School at Fort Belvoir, Virginia to study cartographic drafting (something to do with map making).

I arrived at Fort Belvoir on one of the hottest days ever for that area. It was so hot we were allowed to take off our khaki shirts and wear a t-shirt. It was claimed that this had never been done before. The map drafting course was ten weeks long. We learned how topographic maps were made and skills such as: how to draw roads, place labels and make nice-looking letters. I finished tenth out of thirty students which was a wonder since many of the students had civilian art experience, and I had always finished near the bottom of the class in cursive writing. The rest of the summer didn't seem so hot but seemed to go on forever. I even found a girlfriend (my first ever). She was British and worked as a nanny for a diplomatic family. I went into Washington, DC on most weekends, either to see her or to drink beer with my buddies.

At the end of the training, I and several of my fellow students were scheduled to be sent to Fort Sheridan, near Chicago, where one of the Army's map-making groups was located. When we arrived, we were told that the map-making group was being disbanded and we would be assigned to a "replacement company" until they could figure out what to do with us. This company was used as a temporary home for soldiers being discharged and others who needed to be stashed for short periods of time. We spent six weeks there with nothing to do, although it seemed much longer. We were considered "fair game" for various tedious make work projects and we reacted by "disappearing" right after breakfast and hiding out at various locations on the base. The duty sergeant took it upon himself to hunt us down whenever possible and put us to work at some usually pointless task. The highlight of this whole time was when five of us were selected as an honor guard for a former soldier who had just passed. At this time, the relatives of any former soldier had the right to request a military funeral. We fired a three round salute and stood at attention as a bugler played "taps." Afterwards, the relatives of the deceased invited us to their house where we drank many shots of whiskey in honor of the unknown (to us) soldier.

Orders sending me to Sixth Army headquarters at the Presidio of San Francisco finally came through the first week of November. I scheduled a short leave to see my family in South Dakota. At the end of my leave, a fierce, early blizzard came down from Canada. I left on the Western Airlines DC-3 just before the full fury of the storm hit. This airplane stopped at numerous small towns along the way and it took all day to get to Denver. Denver was also getting the storm with high winds, snow and zero degree temperatures. In late evening, I was on a United Airlines DC-6 heading for San Francisco. However, the airplane had to return to the terminal twice for mechanical problems and didn't arrive until early morning in the "city by the bay." Outside, the sun was shining and it was 60 degrees. I thought I had found paradise.

I was assigned to one of the companies in the 30th Engineer Group located in an area of the Presidio known as Fort Winfield Scott. It was a beautiful hillside setting with a dozen large barracks buildings encircling a large green and facing the Golden Gate Bridge that was less than a mile away. Our main activity was to go to "the plant" each day and work at making topographic maps.

Life in this part of the Army was nothing like what one sees on television or in the movies. Most of the enlisted men had been drafted and many had successful careers in civilian life. They were angry about being in the Army and felt they were intellectually superior to the sergeants who were in charge. They were always looking for ways to torment the career soldiers. On the other hand, the career soldiers seemed to have no idea who they were dealing with and treated us like we were a bunch of illiterate country bumpkins. The song "Working on the Chain Gang" was popular at the time and so we whistled and sang this tune as we marched to work in the morning. This annoyed the sergeants greatly but it took several days for them to come up with a way to stop it. We were periodically called upon to parade in formation around the green. The captain would march us on these occasions and yell "count off," expecting his 100 plus troops to make a great noise, but we would only pretend to shout out the numbers with our jaws opening wide and emphasizing the count but emitting no sound. This drove the captain absolutely nuts.

I was chosen to be the unit's aerial photographer and flew in the back seat of a Twin Beechcraft that was the pride of the group's flight detachment, which also flew a Cessna and several small helicopters. We went on a mapping assignment to the Army's Yuma Test Station, and I flew a number of missions photographing the boundary from 15000 ft.

As my two year hitch was coming to an end, I found out that there was an "early release" program for those who wanted to return to civilian life and resume their college educations. One could get out 90 days early. I also found out that there were Army tests that one could take and get college credit. I took and passed the test for the first year of college. It took only a single afternoon to take the test, but my college gave me 20 semester hours of credit. What a deal! I applied for the early release and got out 10 weeks early.

I returned to the University of South Dakota to resume my studies. This time I decided to change my major to physics. This decision was influenced to a large extent by my time in the Army where I was, if not really an engineer, associated with engineering. The university did not offer a course of study in engineering, but in my view, physics was pretty close. Also, my disenchantment with the sciences caused by my high school experiences had faded. The head of the Physics Department took an interest in me and advised that I sign up for a seminar that he would teach. I was the only student so it provided a lot of one-on-one exposure with the head guy. I didn't understand the special treatment, but it seemed like I was off to a good start. I also took a course in Algebra. This course ran for one semester and covered all the algebra that had been taught in two years in high school and also included trigonometry. It was relatively easy, which made me wonder if my added maturity was a factor, and if perhaps teaching algebra to high school students (other than a few basic concepts), was a waste of time.

I was soon faced with a major problem. I hadn't saved much money while in the Army and I was in financial trouble again. I wrote to my dear old grandmother again and she came through. I made it through the semester and found a summer job working on the assembly line of a large bakery in Sioux City, IA. At the end of the summer, it was clear that I didn't have enough money to return to school full time even if I could find a part time job. I was able to keep my full time job at the bakery and enrolled half-time at the college. I had saved some money by the start of the second semester and then enrolled full time. My older sister, Lois, who had graduated from college the previous year, was now gainfully employed teaching English in a small town in Minnesota and provided some financial assistance.

Unfortunately for me, my physics professor had retired at the end of the semester and my preferential treatment came to a sudden end. The new head of the department was a real character. I had come to expect professors to look a bit strange and behave in odd ways, and this one fit the pattern. One day he announced to the class that the Physics Department was not yet at the level of MIT, but it was going to get there and even be better than MIT. I didn't know at the time what MIT was, but still it seemed to be a rash statement. In later years, I would realize that the claim that his department would eventually achieve elite status was delusional. He taught Physics 101 and for the second semester, gave me a grade of "C." I knew I had earned a "B" and confronted him. His excuse was that I was capable of getting an "A" but hadn't worked hard enough. Where to start? I made a mental note that we would soon be parting company.

The country was in a recession that summer in the second term of Eisenhower's presidency. I returned to my hometown to find summer work. Jobs were scarce, but I managed to find a number of minimum wage jobs such as: shocking oats in the hot sun, loading alfalfa bales, mixing and toting plaster, painting the house my parents were renting and also painting a farmer's barn. My sister, Lois, was home for part of the summer vacation from her teaching job and generously offered to buy a trailer for my brother, Bruce, and I to live in while attending the university. Bruce was also studying physics and was a year behind me. Living in the trailer greatly reduced our two biggest expenses, rent and eating in restaurants. I rewarded Lois by teaching her how to drive and letting her drive my car around town while I was working. I had saved some money over the summer, but it was not nearly enough for the next school year. I was able to find part time work as a reupholsterer's apprentice. It paid minimum wage (75 cents per hour), but I learned quickly and was soon given a raise to 90 cents per hour. There was the temptation to work more hours for more income, which cut into my study time and hurt my grade point average.

I had taken about the same number of courses in mathematics as in physics so it was easy to switch majors. The Mathematics Department was headed by a very reasonable fellow who taught most of the upper level courses himself. We got along well and my grades reflected my actual performance. I took a few more physics courses and performed well in these. There was no backlash from the head of the physics department who seemed to have lost interest in me.

Finances continued to be tight, but with part time work and some help from my sister, I made it through and graduated after summer school in 1960. Job placement assistance from the university was of little help. The companies that did come to campus were from the surrounding area – Minneapolis and Chicago were heavily represented. I had no desire to live and work in the Midwest with the bitter cold winters and sweltering summers. I'd had enough of that. I came to the conclusion that graduate school was a reasonable option and wrote to several schools. The mathematics department in one of the large state colleges in Oregon responded, offering me a position as a graduate assistant, and so I headed west for the fall quarter.

In Oregon, first year college students must pass an entrance exam in algebra in order to take the college level algebra course that was a requirement for most majors. Those who scored low on the exam had to take one or two courses in remedial algebra. I can understand that the state wanted to accommodate the parents and allow the taxpayers to send their children to a state college even if they had not learned any algebra during their high school years, but it might have been kinder to not admit them in the first place. My job was to teach two algebra courses, one basic and one pre-basic. My agreement with the school was to teach one class in my first quarter there, but they were short-handed so I had to teach two different levels while pursuing my own studies. It soon became clear why these students had been unsuccessful in learning algebra in high school. They were, for the most part, unmotivated and could not manage to show any interest in the subject. Their parents were putting out a lot of money to send them to college, and most of them were not going to make it.

The best way to learn something is to teach it to someone else. As for me, I had never gotten excited about algebra, but I now began to really appreciate the subject and developed an intuitive grasp of it, which in later years would help me develop many useful applications and tools using relatively simple algebra techniques.

North American Aviation – Autonetics Division

TEACHING TWO DIFFERENT classes took a lot of my time, and it was difficult to find enough time for my own studies. Also, the wage I received had seemed adequate at first, but I found that I was barely getting by. One day, a notice appeared on the bulletin board saying that a recruiter from the North American Aviation Co. would be on campus to interview interested students. I scheduled an interview and had a pleasant chat with the company recruiter. I didn't know much about North American except that they had built the P-51 Mustang fighter plane, probably the best fighter of World War II. A few days later, I received a letter from the company offering me a job as a research engineer with the Autonetics Division of the company for the astounding salary of $7000 per year (equivalent to about $60000 in today's inflation adjusted dollars). I accepted the offer and left for Downey, California at the end of the quarter.

Downey was a pleasant middle class community located in the endless sprawl that extends to the east of Los Angeles.

I had no automobile but I rented an apartment close to the plant and was able to walk to work. I began to save money and, within a few weeks, purchased some basic transportation. Things were looking up.

My first encounter with my new work environment was an eye-opener. While waiting to talk with my new boss, I was handed off to a senior engineer. He quizzed me on my basic knowledge of simple (to him) concepts, things I had never heard of. He must have thought, "Who hired this guy?"

The group I had joined was responsible for evaluating the accuracy of the Minuteman I guidance system, the N10, which the company was building. I was assigned to a project managed by Dr. M, a young PhD, recently graduated from a state university in the Midwest. The Minuteman missile was being deployed in silos for the most part, but there was a variant that would be placed on railroad cars and move along railroad tracks in the upper mid-western part of the country. These missiles would depend on a gyrocompass for north alignment that was essential for missile targeting. This component, called the North Seeking Gyro (NSG), was being developed by a subcontractor. My new boss was evaluating the NSG for sensitivity to vibration in the environment. It soon became clear that my mathematics and physics coursework had not prepared me to do any useful work. Dr. M seemed to understand this and assigned me to do some curve fitting in the frequency domain, a task which seemed to have little, if any, value. He took me with him on visits to the subcontractor, and although I could not make any useful contribution, I got to observe the process of subcontractor management. The vibration environment was thought to induce a gyro drift known as theta-phi-dot, a combination of a misalignment angle and a drift rate. Dr. M was concerned that this effect would cause a serious degradation in gyrocompass accuracy. There was a PhD representing the subcontractor, a Pakistani with a long hard to pronounce name, (Abi _ _ _), who didn't believe in the theta-phi-dot effect and had an ongoing argument with my boss over this. Dr. M was an intense, serious, no-nonsense person, and I was sure that he was right.

A few weeks later, a test version of the NSG was delivered to our laboratory. It was soon up and running and subjected to a number of tests including one in which the base was rotated in increments through 360 deg. The device was supposed to be able to find North within seconds of arc. The test showed errors of up to 4 minutes of arc that were cyclical at twice the rotation angle. This puzzled everyone and the best minds in the organization were put to work on this problem. It was eventually determined that the rotation bearing for the gyrocompass assembly was misaligned in level and that had led to the error. In the end, it didn't matter. The contract was soon canceled when the Mobile Minuteman concept was terminated in favor of an expanded silo basing approach.

The group I was in had an obligation to produce a document containing an "error budget" that showed the "miss distance" contribution of the N10 inertial guidance system due to fabrication tolerances and output noise levels of the components, periodic degradation of calibrated instrument parameters and mechanization algorithm approximations. In essence, any and all contributions to miss distance that could be tied to the N10 were to be identified and "estimated." The analysts and scientists in the group had so far identified some 40 or so such error sources that defined the accuracy (or rather, the inaccuracy) of the system. This document was to be produced and transmitted to the US Air Force quarterly, but it was such an onerous task that it was done annually at best. As pressure built up over time to have this document produced, someone within the group would be tapped to take on this task. The designee had always been some high-powered PhD. After 2 or 3 iterations of producing this document, it may have been decided that the process was well enough defined so that someone of lesser experience and education could accomplish the task. Anyway, I was given the job. I began by making the rounds and interviewing the managers of the gyro, accelerometer and level sensor groups. My visits must have been cleared in advance because all of them were helpful and supplied the data I needed. I went down the list of error sources and determined who I needed to contact to get the latest data. I had to study the error model documentation for each error source and fold the new data into the calculations of miss distance. I, of course, learned a lot about the N10 system and all of the analyses that had been done to determine the accuracy of the N10. In the process, I was exposed to a new discipline, probability and statistics, which is a fundamental concept used in error analysis. Most of the errors were modeled by starting with the nominal value of the parameter and adding an amount of noise (i.e. randomness) that was from a normal distribution. I thought it a bit strange that I had never heard of probability and statistics in my academic work. My effort was successful and the resulting document was sent out to the appropriate authorities and interested parties. In fact, it was so successful that producing this document every quarter as called for in the contract became my regular responsibility. It is ironic that this document that had seldom been produced and always late was now expected by the deadline date and was the subject of numerous phone calls from our contracting group in the days preceding its scheduled release asking for assurance that it would, in fact, be on time.

It was not at all clear to me why I was chosen for this task of working with other groups to acquire data and update the error budget. Perhaps I was selected because I seemed to be articulate and able to talk to other people and also because I was reasonably well-groomed and normal looking. Engineers do have the reputation of being introverted and not good at communicating with others. It is true that quite a few engineers fit into this category, but I also noticed that many of the engineers in the group made an effort to increase their social skills and became quite proficient in their ability to communicate as they advanced in their careers. Making presentations of one's work is more or less a requirement and being able to do that well is something we all worked on.

I now knew what the accuracy of the N10 guidance systems was, but it was only a part of the overall system. There were other error sources including target location uncertainty, gravity anomalies, atmospheric density variations, reentry vehicle dynamics and others. I was curious to know what the overall accuracy of the missile was thought to be. This would, of course, be classified "secret" as were the N10 errors that I had been computing. I asked around, but no one in the group seemed to know, and no one had any knowledge about where such information might be found. Accuracies were specified as CEP (circular error probable) which is defined as the radius of a circle in which there is a 50% probability that any given missile would fall. I gave up on finding this information until one Sunday when I was reading the LA Times, and there it was, an article on the Minuteman reporting that the missile was expected to fall within one mile of the target. So much for "secret."

There was a bridge game that took place every day at lunch-time. I had learned to play bridge during my last year in college. I participated in this game whenever a slot was available and found that my bridge playing skills were almost as good as the best noontime players. Someone in the company had a connection to The Cooper Union, an engineering school in Brooklyn, New York. He had recruited five or six engineers from this school. They were like me young, single and interested in playing bridge. They were also Jewish and seemed to be articulate, intelligent and generally very good bridge players. We became friends. I had little, if any, knowledge of contemporary Jewish culture, and tried hard to not say anything insensitive although I occasionally did. They were, however, tolerant and cut me some slack. They introduced me to New York delicacies such as pastrami and corned beef sandwiches. I also learned to appreciate Jewish comedians like Shelly Berman and even Lenny Bruce. Al, from this group, claimed to have some experience playing duplicate bridge and suggested that we go to the company recreation center in Inglewood and play in the weekly game. I don't think I had ever strained my brain as hard as that first game of duplicate. My mind continued to race for a couple of hours after the game. We were thrilled to come in fourth out of ten tables and earn .05 master points. Al thought of himself as somewhat of an expert on bridge and lectured me constantly on the proper bids and how to play the hand. This was a bit annoying but I decided to put up with it because I had much to learn. A

week or so later, Ira, another of my new friends, and I went to
the company recreation center in Downey and played in the
weekly game. We were both quite surprised when we found
out that we had come in first while sitting East-West. This was
clearly worth another try, and so we played the next week.
What a shock! We won again. We were the winners for four
weeks in a row. The next week when we showed up to play,
the club director was waiting for us. "Make them sit North-
South!" The players seated North-South are traditionally the
better players, and they stay at the same table all night while
the East-West players move after playing 2 or 3 hands at the
same table. We finished somewhere in the middle of the pack
that night, but it had been a good run.

A few weeks later on a Saturday morning, Al and I went to
Santa Monica where there was a weekend bridge tournament. The morning's event was the "Charity Game," probably
so named because the entrance fees were donated to a local
charity. We had a very good outing and came in first. This
event seemed to have some prestige associated with it, and
we were each presented with an attractive plaque in recognition of our victory. A year or so later, both of my usual bridge
partners decided to return to school to work on advanced degrees, so I began looking for a new bridge partner. Our work
area was divided into four man offices, and one of my of-
ficemates, Fred, claimed to be an accomplished bridge player
although he had not played much "duplicate." There always
seemed to be a weekend bridge tournament in one of the
many communities that comprise the LA area. We began going to these tournaments and clicked immediately. We won
the "George Raft Pairs" (all events were named for Hollywood
notables that year) at "Bridge Week," held at the Ambassador

Hotel. We usually played in a "side game" at the smaller tournaments and won most of the time. Winners were awarded 1.5 "master points" and a gift certificate worth about $5. One Friday afternoon, we decided to duck out of work and play in the "Men's Pairs" event at the South Gate Sectional Tournament. The "Men's Pairs" is a prestigious event, and it attracted several high-caliber players from the local area and two or three nationally ranked players. We played a bidding system similar to the one described in the Alfred Sheinwold classic, "5 Weeks To Winning Bridge," and only made an opening bid in a major suit that was at least 5 cards in length. For some reason the benefits of opening with only 4 cards in a major suit had caught our attention and we decided to experiment with this approach for this tournament. The result was that we ended up with the bid on several hands with a 4-3 fit in hearts or spades. I played 3 or 4 of these hands at game and had never worked so hard at card counting. We won the whole thing by a big margin and were awarded 4 master points, our biggest haul to date. I think our success was due to making several games with a short trump suit while everyone else was in 3 no-trump. Strangely enough, we never played "4 card majors" again, probably because of the extra effort and stress required to do so. There is no such thing as a perfect bridge partner, and although generally good in most respects, Fred had one annoying quirk. Bridge is a game of percentages, and the objective is to know these and make the bid or play that has the best chance of working. Simple enough? On occasion, Fred would seem to be channeling a message from the beyond and make some strange bid, ignoring the obvious and odds-on choice. This usually got us in trouble, resulting in a low score. No amount of brow beating or harsh criticism could break him of the habit. He clearly knew better.

One of the most memorable hands I ever played occurred one evening when I was partnered with a young female player, an occasional partner. She would on occasion open with a weak third hand, a practice I had heard of but one that I avoided and failed to recognize this time. Consequently, we wound up in 3 no-trump with slightly more than half the high card points. We were playing against Erik P., the best player on the West Coast, and Helen P., another top notch player. My only hope was to bring in the 5-3 diamond suit. I took my ace and led to the king/jack on the board for the finesse. As I did so, I watched the facial expressions of both my opponents, which seemed a bit too serious for a simple finesse (that they fully expected me to take). I pretended to think for a moment to get a better read on Erik, who seemed a bit too tense, and then played the king, dropping his queen and going on to make the contract. I'm sure they got a bottom board on that one. It probably meant nothing to them but was something I'll never forget.

In the next two years, I played with a number of different partners and was proud of the fact that I had been able to come in first (at least once) with all of them, although this string was eventually broken. My bridge playing career ended after less than 4 years. My regular bridge partner took an assignment with a different division of the company and left the LA area. I found a few other partners, but my interest in bridge began to wane as I found other interests. During my bridge career, I had accumulated over 50 master points, making me a "National Master."

In October, there was an annual event sponsored by the LA Times newspaper called the Riverside Grand Prix. In 1962, I went with a colleague from work to this event. Some of the top sports and grand prix drivers in the world showed up for the races. There were a number of races for different classes of sports cars, but the final race on Sunday was for unlimited modified sports cars and drew all the top drivers. Dan Gurney, a local favorite, was enjoying a big lead with his Lotus 19 until a part in the throttle linkage broke. The race was won by Roger Penske who was at the time a virtual unknown. There was some controversy over his win because he was driving a single seat modified Formula 1 car. I think they changed the rules after that to redefine "sports car." Some amateur drivers were allowed in the race to run with the pros. A favorite car for many of the amateurs was the Lotus 23, a sleek looking machine although less powerful that the Lotus 19. Near the end of the race, we were watching at turn 9, the last turn before the finish line. Pat Pigott in his Lotus 23 came around the turn a bit wide and seemed to slip under the guard rail and hit a post head-on. The car virtually exploded on impact with parts flying everywhere. All that was left was the seat with the driver's motionless body. He later died in the hospital. It was a horrible tragedy.

I was driving a TR-3 sports car then, and several of my friends also had sports cars. I recruited them to go with me to the races the next year. We got our tickets early and nailed down some good seats in Turn 2. That was the year of the King Cobra. Dave McDonald was a popular local driver and led the Cobra Team. He won handily and thrilled the crowd with the way he slid the car through Turn 2. The story does not have a happy ending. McDonald drove in the next Indy 500 in a Mickey Thompson car that many of the drivers thought was dangerous. He crashed on the second lap in a fiery collision with the wall and died in the hospital. The popular and beloved driver, Eddie Sachs, was blinded by the smoke and ran into McDonald's disabled car. He was also killed.

The next year we recruited a large group from the office and took over the top rows of the bleachers at Turn 4. We brought our beer coolers along and enjoyed a sunny day of racing. The race was won that year by Parnelli Jones driving a Cooper King Cobra.

Our group was on the hook to produce an "engineering manual" that described all of the analyses that had gone into characterizing the "error sources" that contributed to the miss distance of the N10 inertial guidance system. The pilot document had been produced earlier, but regular revisions were also required because new analyses were constantly being generated and old versions revised. As usual, this document was overdue. I was assigned, along with another engineer of similar limited experience, to produce the required revision. We got very little direction from our boss and never had any status or planning meetings. This did not seem strange to me at the time since I knew nothing about such things. We divided up the work between the two of us and worked at our own pace. The only encouragement that we received was an occasional comment about how it was taking us a long time to complete the task. This was actually a very good learning experience as we were forced to understand a variety of some really complicated analyses. We eventually finished the task and produced a high quality document.

An important part of error analysis involves the computation of partial derivatives of impact error (downrange and crossrange) with respect to the velocity errors (on the x,y & z axes) at thrust termination. In the early days, these were calculated analytically, but with good simulations of missile flight available, these partials could be easily and accurately calculated. The method was to first fly the nominal trajectory to impact, and then fly the missile again with 10 fps (for example) added to the x axis. Next calculate the difference in downrange position between the two impacts and divide by the 10 fps. This is the partial derivative of downrange position with respect to x axis velocity. The same approach is used for all of the other (downrange and crossrange wrt position and velocity) partial derivatives. I thought that this was really clever and would find the method useful in many future applications.

North American Aviation's Space and Information Systems Division (S&ISD) was awarded the contract for the Apollo Command/Service Module and various system development functions in 1961. S&ISD launched a frantic search for talent and began raiding Autonetics for bodies, with the apparent approval of the corporation. Several of our PhD's signed on and transferred to S&ISD. I visited Dr. H. at his new location. He was working on the Earth rendezvous problem where the moon stage would be launched into orbit in pieces and assembled there before heading to the moon. This concept, of course, lost out to putting everything on one big booster rocket. S&ISD grew rapidly and eventually laid claim to our facilities in Downey, forcing us to move to another NAA location in Anaheim, CA.

The N10 guidance system would be subjected to accelerations of up to 8 or 9 g's during the missile boost phase. Several of the error sources for the system are driven by these accelerations that are constantly changing along the missile trajectory. Computing the miss distance attributable to such errors was a complex calculation and required, well, a computer. The company had a large data center equipped with several IBM 7090 and 7094 computers which were the latest available. The engineers were supported by a group of computer programmers who were mostly young women and who tended to be fairly new to the field. One of the engineers in the group had been given the job of computing the errors due to acceleration dependent gyro drift during the missile boost phase. This required that computations be made as often as several times per second during a simulated missile boost. He and his programmer were often seen working late and shuffling back and forth to the data center. Each computer run produced a stack of printout several inches high. There appeared to be only a few lines printed on each page with a page being printed for every computational cycle. In time, the engineer's office was filled floor to ceiling with stacks of the results. Computer programming was a relatively new discipline at the time, and all of that wasted paper was probably a byproduct of the learning curve.

The use of programmers was felt to be inconvenient by many of the engineers, and they began to learn to write their own programs. FORTRAN was the language in use. The available reference that described how to assemble a FORTRAN program to run on one of the computers was issued by the IBM Corp. and proved to be inscrutable for many of those who tried to use it. I initially tried this route myself without any luck. I finally asked another engineer who had been successful to show me how to write a simple program with nothing more than a format statement to print out some data. This succeeded and I was on my way, adding to and expanding the capability of my original effort. One of the engineers took on the task of writing a FORTRAN program that included most of the complex error computations and printed out the results as a concise summary. One of the inputs was a boost phase acceleration history. A number of boost trajectory options had been created and with that, error calculations could now be made for a variety of missions.

The company had an interesting way of dealing with problem employees. This included poor performers and possibly those who had behaved in an inappropriate way, which may have included moral or legal offenses. On three occasions that I know of, the employee simply disappeared without a trace. As the rest of us arrived for work the next day, we found that the subject individual's office had been cleaned out leaving no indication that the space had been occupied the day before. No one would admit to having any idea as to what had happened. One case of non-performance stands out. Dr. N, a newly minted PhD, joined the group. He developed the habit of entertaining himself by stopping by every day to chat with me and a number of other engineers and seemed to spend his entire day pursuing this activity. He never appeared to be doing any useful work. I made a concerted effort to discourage him from wasting my time, but he continued to make his rounds. He left for a few days to travel somewhere (he said) and never returned.

Working conditions were somewhat challenging, especially for the junior engineers. At the Anaheim facility, the senior engineers had cubicles of about 45 square feet and containing a desk, a chair, a visitor's chair and a small bookcase. The junior engineers were in an open area containing a dozen or so desks. It was often noisy with many conversations taking place and people going in and out, making it difficult to concentrate. About forty of us were housed in a large room with a mix of cubicles and open areas. A lot of the engineers complained about the noise, resulting in a visit by the facilities people. They looked around and took some measurements. In time, they gave us their recommendation, which was that we should observe library silence. We gave them credit for having a warped sense of humor. Shortly after that (and maybe in response to our complaints), we moved to another building where the work area was divided into four man offices for all of us, a much better setup. Still, I was often bothered by the noise and found it hard to get my work done but worked at tuning out the distractions. I came across an LP record on self-hypnosis and used it to help my concentration. I think it helped a lot because I was able to spend most of my time in the office doing useful work. When my performance review came around after four years with the company, my boss told me that because of my superior performance, my salary would be increased to put me in the top ten percent of engineers. I was too naïve to ask the right questions like: How many dollars would that be? How long would this take and what would happen in the next year? Etc. I found it a little hard to believe that just by spending my time doing productive work and getting things done I stood out, but that was apparently so.

The N10 guidance system was being replaced with the N16 for the next generation of the missile, the Minuteman II. The new system was externally gimbaled, which eliminated some of the problems of the internally gimbaled N10, such as pitch post bending under acceleration, which caused a major misalignment of the accelerometer package. The system was composed of a light metal shell with cutouts and plates for mounting the inertial instruments and was supported by external gimbals. A new computer made extensive use of integrated circuits and was contained in a relatively small box instead of being spread throughout the guidance compartment and virtually occupying every spare bit of space as with the N10 on Minuteman I. After having produced an "error budget" for the N16, I was asked to come up with an estimate of what improvements in performance might be possible in the next year or two. I visited my contacts in the gyro, accelerometer and alignment groups and put together a summary of their estimates and the expected total performance improvement. I published my results in a memo addressed to all of the management of the groups that would be involved. I thought my job was done until I was summoned to the second level boss's office one day. As I entered the room, I noticed a collection of second and third level managers. They had apparently been reviewing my memo. I was asked a few questions about some of my numbers and was dismissed. Maybe they just wanted to get a look at me. I later learned that my memo and its accuracy predictions were being used as the basis of an accuracy improvement program worth millions of dollars being proposed to the Air Force. I felt somewhat important after this, but I also realized that if I hadn't produced the document, someone else would have.

The guidance system analyses used a lot of different coordinate systems requiring transformation matrices to navigate from one to the other. Some calculations involved consecutive rotations about different axes. One of the engineers invented a "matrix wheel" that allowed one to perform rotations of the wheel and view the resulting transformation with the alpha, beta and gamma sine and cosine products showing through the cutouts on the wheel. Everyone became skilled at matrix algebra and computing transformation matrices. This is an ability that I would find highly useful in future assignments.

I had heard it said that, "Mental maturity, if it comes, occurs at about age 27." I had learned a lot in the last few years dealing with people in both work and social situations. I still had a lot to learn. I had for some time worked hard to control these relationships to get what I wanted and to influence events to unfold in my favor. I realized that I was expending a lot of energy in this regard with limited success. One day it occurred to me that trying to control the world was a futile endeavor. In a moment of great mental clarity, I realized that the answer was to accept the world as it was and change my behavior to fit in with it.

I had wanted to learn to play golf for some years, and when I was finally out of school and working, it was time. There were two other engineers in the group, Joe and Lloyd, who were also interested. We arranged for group lessons at the Disneyland driving range and signed up a dozen or so others from our group. After a few lessons, we were ready to try playing and started going to the Disneyland par 3 course located just behind the hotel. Eventually Lloyd and I were brave enough to try a full sized course and went out to the Anaheim Municipal Golf Course. This outing was a disaster. I kept swinging harder and harder after every missed shot and was absolutely exhausted after nine holes. But we persisted and in time, were able to play at an acceptable level. With all the golfers in the group, we had enough players to have a regular mid-week foursome at the Anaheim course, playing early in the morning and making it into work almost on time. There was also a regular weekend foursome at one of the many courses in the area.

In the winter, you could identify the skiers in group by their tan faces. Joe had some skiing experience and pressured Lloyd and me to join him. I had never skied but was willing to try. The best ski area around was Mammoth Mountain in central California, a 7-8 hour drive. We went there for the weekend. I rented skis and began practicing on the rope-tow slope, and by Sunday, I was far enough along to get on the T-bar. We began going to Mammoth to ski every other weekend during the season and eventually decided that the no. 14 chair was the best run. We skied hard and tried to do 20 runs per day. So, we were playing golf in sunny warm weather during the week and enjoying winter on weekends. We ate in restaurants most evenings, barbecued steaks, drank beer and played poker – the normal things that young single guys do. It was quite a life. However, things would soon change.

Opportunities to meet suitable young women were somewhat limited. There was, however, the temptation to date the girls in the office. Looking back, I don't recommend it. If there are problems, interacting at work can be embarrassing and the gossip network can have a field day. Also, pursuing a romance during working hours is a distraction that can have a negative impact on one's performance. There were, however, a couple of workplace romances that worked out well; at least, they ended in marriage. Joe began dating Georgia, a young female engineer from another group, and she joined our snow skiing and waterskiing outings. Lloyd met Sandy at a dance at the

Disneyland Hotel and they began dating. Some of the other guys that were in our extended ski group also began bringing their girlfriends. We stopped going to Mammoth Mountain and moved to nearby June Mountain, which had some less challenging slopes. Our ski trips had evolved and were now a mix of single guys and couples.

I was still driving the Triumph TR3. It was a lot of fun to drive, but the girls I dated didn't appreciate that aspect. I received a lot of complaints; didn't like the way I took the corners, didn't like the way I slammed the door, etc. I somehow thought these were complaints about me and didn't realize that the car may have been the problem. One Friday evening, I attended a dance at the Ambassador Hotel in downtown Los Angeles sponsored by a group called The Alumni Club. I have always been attracted to the prettiest girl in any group, and that night it was a young blond named Carole. We danced and had some conversation. I found it interesting that she also worked for NAA, although in S&ISD (Space and Information Systems Div.), giving us something in common. I was able to get her phone number and called her the next week for a date. She zeroed in immediately on the problem with my transportation. With her tight skirt, it was hard to get in and out of the TR3, and she made that clear. After a couple of dates, things seemed a bit iffy, so I went down to my local Ford dealer. At that time, the new Mustangs were very popular with the younger set, and I decided that was the car for me. I ordered a fastback model in burgundy with a black interior, a great looking car. I took delivery after a few weeks, and it seemed to be a hit with Carole. We began dating a lot, and she joined our ski outings at June Mountain.

When Carole and I first met, she was sharing an apartment with another young lady but later moved into a house owned by a young Hispanic woman. I'll call her Selena. She was quite attractive and had a well-developed chest. In fact, she often seemed to steer the conversation to the subject of how men were always staring at her tits. She had a couple of men "friends." One was an older gentleman (mid-forties) who was married. He often took her to see the LA Dodger's games. Once when he called for Selena, I answered the phone. He seemed to be very upset that a man was answering the phone although he should have remembered that I was often there to see Carole. Another friend was Eddie, who claimed to be an engineer and might have been a good prospect. He used the friend approach so he could stop by and hang out with Selena. I told Carole that he was trying to build up the courage to declare his love and was getting ready to make his move. He soon did, and Selena told him to stop coming around. Both guys used the friends approach, apparently lacking the confidence to be direct and were perhaps intimidated by Selena's well developed chest. The highlight of Selena's encounters with men was an incident that occurred when she was on an airplane and was hit upon by the star of the then popular TV drama "12 O'clock High." He pursued her for a while but the relationship never took off (pun intended), probably because of wife trouble (his, of course). I did not want to give Carole any cause for concern so I made an effort to not get too "chummy" with Selena and never let my gaze linger for too long on her primary asset. Consequently, Selena was not impressed with me and even advised Carole that she could do better. At the time, Selena was dating a bartender who was in the process of getting a divorce (he said). Oh well, one has to love the irony.

There is an evolution in relationships. What happened next should probably come as no surprise. Lloyd and Sandy were the first to get married in the summer of 1965. Shortly thereafter, Joe and Georgia wed. Carole and I married at the end of the year. We began going places as couples with Friday evening dinner a favorite. The ski trips also continued on for some time. We went to the Riverside Grand Prix for the last time in 1965. The race was won by Hap Sharp in a Jim Hall Chaparral race car. It was just Joe and Georgia and Carole and I attending this time. I guess things were changing, and we were all moving on.

The N16 guidance system contract required that an engineering manual be produced as had been the case for the N10. Since my knowledge of the system error models was more comprehensive than that of anyone else, I was assigned the task. I assembled the available documentation, incorporated that, and performed and documented the analyses that were still needed. I also wrote the overall narrative and included results for a number of cases that showed system capability for a range of trajectory options.

In 1960, an electrical engineer by the name of R.E. Kalman wrote a paper entitled "A New Approach to Linear Filtering and Prediction Problems." His method used a series of measurements to continuously refine an estimate of the system state vector and produced a statistically optimal estimate. The value and implications of this landmark development were soon recognized by some of the best brains in our group as its applications to navigation systems were apparent to them. Several of our PhD's began giving seminars on the subject and we all became pupils trying to learn the wonders of Kalman filtering. The Marine Systems group applied this method to improve the accuracy of the SINS (ships inertial navigation system), an Autonetics navigation system being used on nuclear submarines. The method relied heavily on matrix multiplications and was a bit hard to follow, a fact that prevented all but the most persistent from working through the computations. My boss had constructed a computer program to use the method to estimate N16 calibration parameters and had me make computer runs to study various options. My insight as to what was going on was somewhat limited, but it allowed me to claim Kalman filtering experience on my resume, which helped me later on.

The next upgrade to the Minuteman missile was the Minuteman III version. The missile carried three warheads known as multiple independently targetable reentry vehicles (MIRV). I produced error budgets for this system. A lot of hardware and several complete navigation systems for the earlier versions of the Minuteman had been produced by this time and test data were beginning to come in. The decision was made to incorporate this data into an "accuracy estimate summary" based on actual performance rather than theoretical predictions based on analyses. If significant differences occurred in these two measurements/estimates of the same thing, the discrepancies could be investigated and resolved. I now spent most of my time tracking and updating information on the various Minuteman systems. I was in a group called Ballistic Systems Analysis and had gotten a reputation in the larger organization as one who could perform various analysis tasks for system components. People began coming to me requesting that I help them solve various problems. I was flattered by this and generously agreed to help them. For each of these tasks, I was given a slip of paper with a charge number and a number of hours on it. My mistake was that I failed to involve my boss in this. I didn't really have the time to do the work unless I was willing to work evenings and come in on the weekends. If I had brought my boss on board, we could have delegated this work to junior engineers, who, with a little help from me, could have done the work. As it happened, the little slips of paper piled up, and I never got around to any of it.

One day, I stopped by to chat with the second level boss's secretary. She told me that they were interviewing some PhD, but they couldn't decide on whether to make him an offer. One concern was that he had been with a lot of different companies. She mentioned that he had a hard to pronounce name, Abi-something. I rattled off the name. She said, "Oh, do you know this guy?" I said, "Not really, but I know someone who does." I referred her to Dr. M, who had been my first lead engineer on the NSG project. I knew he would relish the opportunity to rule on the qualifications of his former nemesis. Needless to say, the applicant didn't get hired.

I had been with Autonetics for almost six years. When one is young and in that "first job," six years can seem like a long time. The work was getting to be just more of the same. My latest performance review was good, and my boss said that he was looking for a supervisor slot for me. This would be a promotion to first level management. I was somewhat surprised that he would be willing to let me go, but it was the tradition within the organization that high performance engineers would be moved into management. Somehow the old saying that "you lose a good engineer and gain a bad manager" came to mind. My salary had gone up some, but there hadn't been that anticipated big increase. I had reservations about what management promised. Did they really mean it? Was there any guarantee that they would follow through? Did they even remember what they had said?

My new wife, Carole, had transferred to the Autonetics Division, and we were now even working in the same building. We had a pretty good life. I was, however, concerned about how living in the Los Angeles basin would affect my health long term. The air quality was good most of the time, or so it seemed, but there were many days when it was socked in with low hanging smog that couldn't have been good to breathe. Some days in the afternoon, one could see a brown cloud advancing inland from the coast as the on-shore breeze began to blow. In the evening, the cloud would blow out to Riverside leaving the LA basin clear, but it would blow back in during the early morning hours. With the health concerns and the fact that my work assignment had become routine and that my career didn't seem to be taking off, I thought it might be interesting to work for another company. I wanted to broaden my experience. So far I had only worked on the Minuteman guidance system. I made the decision to leave Autonetics, and having done so, there was really no turning back. My exit interview with the second level boss was somewhat surprising. I learned that I was considered to be the number one (or most valuable) engineer in the Ballistic Systems Analysis group. If I had known that earlier, it might have been a factor in my decision. In retrospect, even a vague plan with some detail as to where I was headed in the company and a significant salary bump would have impressed me. Unfortunately, it was too little, too late.

The Avco Corporation

THE AVCO CORPORATION had been in the news recently with their development of the Apollo heat shield, and they were advertising for engineers to work on a reentry vehicle contract. After a phone interview, they scheduled a visit for me to meet with them at their location in Wilmington, Massachusetts. It seemed like an interesting change and an opportunity to expand my expertise, so I accepted their offer. Carole had never lived anywhere but Los Angeles. It would be an exciting adventure for her. We moved to Massachusetts in January, probably not the best time to do so, but it seemed like a relatively small sacrifice. We rented an apartment in a picturesque wooded setting in the small town of North Andover.

On my first day at work, as my new boss and I surveyed the work area, he seemed to be almost talking to himself as he said, "I don't know what I'm going to do with you." The reentry vehicle contract that I had been hired to work on had been canceled by the time I arrived. Fortunately, the department had some R&D funds, and in the next few months, I was put to work on a number of different research projects.

The work area consisted of one large room with offices around the periphery and a smaller open area for the senior engineers, where I was placed. One of the first things that I noticed about our work area was how incredibly filthy the floors were. They looked as though they had not been cleaned for years. I thought this was a very unhealthy situation and asked my new leader if we could have them cleaned. After some discussion and his assurance that nothing could be done, he conceded that it would be OK for me to contact facilities/maintenance if I wanted to waste my time doing that. I decided to write a memo to my boss complaining about the unsanitary conditions and possible negative implications affecting my health and that of my colleagues. I may have also mentioned the negative effect on workplace moral caused by having to work in such conditions. I, of course, cc'd his boss, as well as, facilities/maintenance. It was probably not after the next weekend but maybe the following weekend that we came in to work to find that the floors had been cleaned, waxed and buffed. The power of putting one's complaint in writing and creating a permanent record was clearly demonstrated. I had no idea why the floors had been allowed to become so filthy or why no one (before me) would have put some pressure on the maintenance people to do their job.

My first project was to recreate the engineering manual that I had written for the N16 inertial navigation system. I was able to produce this from memory and included everything but the numbers which would have made it a secret document, and besides, the numbers were always changing. However, some of the information was still classified, though how classified was an issue. It seemed like I was free to assign any classification I thought appropriate. I chose to make it "confidential," which would limit distribution and prevent it from falling into careless hands (I hoped.) and, of course, protect me to some degree from being guilty of compromising classified information.

Strap-down guidance systems were a relatively new approach to inertial guidance that eliminated the need for gimbals. The gyros are fixed to the case containing the guidance system and torqued at the appropriate rates to allow the gyro wheels to follow the rotations of the case. This approach was appropriate for smaller less accurate applications such as some reentry vehicles. I was given the funding to pursue an analysis of the accuracy that could be attained by a strap-down system. I developed error models for the accelerometers and gyro drift effects. My approach was to calculate errors in a case referenced coordinate system and transform these back into an earth fixed computational coordinate system where they would be summed. The transformation matrix was updated at a fairly high rate to reflect the rotation of the case as indicated by the rates at which the gyros were being torqued. I produced some results for a few representative trajectories and felt that I had demonstrated the approach. I wrote a memo and distributed it to a list of interested engineers and managers.

I was now known for having some level of expertise with strap-down systems, which led to a rather bizarre episode. A project manager from another corporate location stopped by one day looking for some help with a difficult problem. He had been awarded a small contract by the US Air Force to perform some tasks related to a reentry vehicle guided by a strap-down inertial navigation system. The vehicle was onboard a spy satellite that would photograph areas of the Soviet Union and then drop a guided capsule containing the data as it passed over US territory. The problem was that this manager had performed little if any of the work and time was running out. My boss selected me to help this guy out. As an engineer who is working at the grass roots technical level, it didn't seem like I was the right choice. Looking back, what was needed was a management consultant to look at the contract requirements, review the task breakdown structure (if there was one), identify the deliverables that these tasks would produce to show that the work was done and that contract requirements were met, estimate the resources needed to complete the work and establish a realistic timeline for the work to be completed. The only other thing for the consultant to do would be to advise that this manager be fired. One can only speculate on why this was not done. Perhaps my management didn't know what to do or they simply wrote this guy off as a lost cause, their only concession being providing me to help him. We started out with me trying to explain to him what I did. The manager seemed to think that what he needed was some magical computer program that could crank out some data that would satisfy the contract requirements. I tried to convince him that no such program existed and that it would take some time to create anything that might help. After going around on this for some time, he said that we should go to Hartford, Conn. He had once worked for United Aircraft in Hartford and knew some people there who might be able to help him, or so he thought. We got into his rental car and drove the two plus hours to the UA facility. He had arranged a meeting that included a couple of mid-level managers and a retired 3 star General. I didn't recognize the General's name, but I was told that was because he had been in Army intelligence, and his name, as well as his activities, had been "hush-hush." Our conversation meandered but included the hope that UA might have some magical program that could generate the kind of output that my manager buddy could use. We, of course, came up empty but succeeded in killing the rest of the afternoon. As we drove out of Hartford on our way back, we passed a bar that my guy seemed to know well. "Let's stop for a drink," he said. I said no and that I really needed to get home. Well, he was driving so we stopped at the bar anyway, and he ordered a drink. I refused to join him. I was afraid that if I started drinking with this guy, we would never get out of there. I wondered if booze might be the reason he was having job performance problems. I was able to get him back into the car after two drinks although he wanted to stay longer. We made it back to Avco safely and parted ways. I never saw him or heard from him again.

After six or seven months with the company, I was transferred into a different group. It was the group I probably should have been with from the start as they were doing various analysis tasks associated with the Minuteman missile. Minuteman III uses a post boost vehicle or "bus" to deposit its three reentry vehicles and a number of penetration aids or decoys on the desired trajectories. Avco had some interest in how the bus worked since they were involved in the reentry vehicle end of it. I offered that I could develop models for the bus and its operation. Although I had only partial detailed knowledge of the bus and its mission, I thought I could come up with a credible design that could do all the maneuvers that the bus needed to do. I spent a few weeks deriving the necessary equations to compute the required velocity changes, the vehicle orientation maneuvers and rocket engine burn physics needed to hit a set of selected targets. The analysis also accounted for maneuvers such as "back away" and established a time-line for each bus action. I produced a memo with my results, which my manager reviewed. He thought it should be turned into a computer program. A computer programmer was assigned to help me in this effort. Building a program for which there was no current contract need, but perhaps something in the future, didn't seem like a cost effective strategy, but the company seemed to want to build up its library of computer programs.

Massachusetts has a culture that is quite different from any place I had lived before. In particular, the accent was a challenge at times. When I answered the phone at work one day, the caller asked for Mah-dee Cully. I assured him that no such person worked here. I was wrong. He was calling for Marty Curly. The natives had never learned to say their r's except when they chose to put one at the end of every word ending in the letter a. I decided to have a little fun with my programmer. I was using the Greek letters alpha and beta to represent angles, and he insisted on calling them alpher and bayter. I created two new angle variables, alpha-r and beta-r. He was never able to pronounce or keep them straight and eventually just stopped talking about them. In the end, we produced a pretty good program which allowed one to input a launch point and a set of targets and let the bus fly around and drop off the RV's and penetration aids on trajectories that hit the chosen targets.

Carole was pregnant and stayed home most of the time to await the birth of our baby expected in July. There were several young families living in our apartment complex, which may have contained as many as 30 three-story apartment buildings each containing 24 separate apartments. Carole made friends with a number of the wives and we had an active social life getting together with the other couples. We were invited over one evening by (I'll call them Don and Mary) one of the couples for dinner to be followed by a friendly poker game. As the poker game progressed, Don and Mary found themselves several dollars ahead. I didn't care because we all seemed to be enjoying the game, and at best, only a few dollars were at stake. However, Don began to gloat and chide us for losing. I found Don's behavior to be offensive and decided that he needed to be taught a lesson. When Carole got a good hand, she could be counted on to bet more, and I could tell just how good it was by the expression on her face. When this happened, I made big raises to build up the pot even though I had nothing. The result was that Carole won some big pots and we ended up the night far ahead. I don't think Don had a clue about what had happened.

My brother, Bruce, had recently moved to the Boston area, having taken a job as a microwave engineer with one of the international telephone companies headquartered in Boston. We had a regular weekend golf game, weather permitting. He helped us explore the local restaurants on weekends. Some were quite good, but the décor in some was strange. Picture a large plain room with a wood floor and colonial style wooden tables and chairs. The intimate dining style popular in California had more appeal for us, but it had yet to make substantial inroads into the Massachusetts culture.

The group I had transferred into had been asked to do a number of analysis tasks but didn't have anyone who knew how to do many of them. I proceeded to work my way through the list and document my results in memos. One of these tasks was to determine the accuracy requirements for the accelerometers and gyros on a data instrumentation package (DIP) that would operate during the post boosts phase of Minuteman test flights. For single RV Minuteman missiles, the RV is popped loose from the booster at thrust termination. There is an attitude control system (ACS) attached to the RV that controls a pitch maneuver followed by a de-pitch maneuver to orient the RV at the proper angle for reentry. The RV is then spun up to stabilize it at the desired orientation. The DIP is attached to the ACS and contains accelerometers and gyros so that it can measure the rates and accelerations experienced during these RV maneuvers. Avco was requested to determine accuracy requirements for these gyros and accelerometers. An engineer had previously been assigned this task and had declared that there was no way to do it. It was important to know accurately the velocity changes that occurred during RV deployment since the velocity mean values could be programmed into the missile guidance software and compensated for. Also, knowing the rates on the RV before it was spun up were important because they would determine the RV coning angle and the atmospheric drag from the mean value could be compensated for. The DIP needed to measure these effects accurately to be of any value. Gyro random drift rates and accelerometer bias errors were the sources that were significant. I proceeded by calculating the error levels for each of these that would result in 100 feet of error at the target. These results could then be scaled to yield a permissible and/or required level of error and then determine the instrument accuracies needed to produce the desired result.

Another task was to develop an algorithm that would maximize separation distances for consecutive chaff canister ejections and minimize chaff wafer rotation rates. I also developed a computer simulation to compute collision probabilities between penetration aids (called gyro-jet reflectors) and the reentry vehicle.

By the summer of 1968, I had been with Avco for a year and a half. I had gotten a raise, which was "decent" and had made a place for myself with lots of work to do. Our son, David Michael, had been born the previous July at the Bon Secours Hospital in nearby Methuen and was now walking around. My wife and I had adjusted to Massachusetts quite well and were thinking that we might stay for a while. Unfortunately, the Defense Department began cutting ballistic missile contracts, and several programs that Avco had been working on were canceled or severely cut back. There was panic in the ranks as engineers were hit by a series of lay-offs. Some senior people were being let go. I didn't pay much attention to this as I was very busy, and management liked my work, but I began to notice that everyone was feeling the stress. One day, I talked with a group of very senior engineers. They all seemed worried about being cut in the next lay-off. I suggested that they might look around for a job with another company. They said, "We are looking, but there is nothing out there." This forced me to think about my own situation and I decided to take my own advice. I answered two ads in the Boston paper and received replies from both. The first was a small company in Santa Barbara called the General Research Corporation (GRC). One of their VP's was in town doing interviews. I met with him, and we had a pleasant conversation. He was hiring for a subcontract the company had with Bell Laboratories in

New Jersey to develop a simulation of the Sentinel missile defense system, a new DoD initiative. The work seemed to be heavy on computer programming but did tap into my knowledge of missiles. They offered to relocate us to Santa Barbara after a year. The other offer was from the Boeing Co. They were interested in the Kalman filter experience that I had cited in my resume. They flew me to Seattle for an interview. I would be working on the development of an aircraft inertial navigation system that used Kalman filtering to make in-flight updates. On my return flight, I changed planes in Chicago and boarded a stretch DC-8 for the last leg of the trip. There were no more than twenty people on the flight. The cutbacks in defense spending seemed to have severely impacted business travel.

Both companies made offers of about the same dollar amount. My wife and I really liked the idea of moving back to California, especially Santa Barbara, but I was wary of going to work for a small company. The Boeing assignment was in Alamogordo, NM. I wanted to hire in at Seattle and be on a temporary assignment in New Mexico, but the company insisted that I hire in at the Alamogordo location. My concern there was that I could be out of a job and in some small desert town when the project ended in a few years.

But with two offers in hand, I decided to ask both of them for more money. What did I have to lose? General Research came through with another healthy increment. Boeing said no. So in the end, it was an easy decision and a good one. Massive lay-offs had already started at Boeing in Seattle.

Carole and I decided to use some of my vacation time and

see Washington, DC while we were on the East Coast. We thought we could drive there and started out early one morning. We hadn't done much planning for the trip but once on the road, realized that it would be a long drive that would eat into our vacation time. We revised our plan as we drove, did a one-eighty and headed for Logan Airport. Back then, there were usually empty seats on the airplanes, and we had no trouble getting a flight. We landed at Dulles Airport and rented a car. This was in early October. The weather was pleasantly warm and the tourist crowds were gone. We were able to park right in front of the Lincoln Memorial. We traveled to Mount Vernon where there was also no waiting. I have a picture of my son playing on the lawn in front of George Washington's house. In all, it was a very enjoyable vacation.

My boss and coworkers gave me a nice going away party at one of my favorite restaurants. I was drinking Manhattans back then. My boss kept them coming, and I realized too late that I'd had too many. I was able to make it to the phone booth and call my wife who came and got me. I've never been able to look another Manhattan in the eye.

General Research Corporation

New Jersey

ON ELECTION DAY in 1968, Carole and I loaded up our two cars with our remaining possessions, having already shipped our furniture to storage in New Jersey. We still had the Mustang and also my brother's new Volvo sedan, which he had entrusted to us while he was on a foreign assignment in Chile. It was a bit late in the day when we left Massachusetts, but we were anxious to get started on our new life in New Jersey. We stopped for the night at a motel in Hartford, CT. My wife and son went to bed early while I stayed up to watch the election returns. As I recall, it was around midnight before they called it for Nixon. The next day we finished our journey and moved into a Howard Johnson Motel on busy Route 46 near the town of Denville. I was scheduled to report for work at my new job on Monday so we had a few days to look around and hopefully find a place to rent. We were shocked to find that there was a universal policy that apartment owners would not rent to anyone with infants or small children. This was not going to be easy.

On Veteran's Day, I reported in for work at the Bell Labs facility in Whippany, NJ where GRC had been allocated some space for the Sentinel Project. GRC was working even though it was a holiday; Bell Labs had the day off. I met the boss and the rest of the crew. The majority of them were staff members from the Santa Barbara facility and had been pressed into service for the contract. The rest were like me, new hires from the East Coast. The SB people had all received financial incentives including generous living allowances and were also allowed to relocate their families to NJ. They could return to SB after one year.

My role was to develop computer models for the Spartan and Sprint missiles. The Spartan was a mid-course interceptor and the Sprint was designated for atmospheric intercepts. We were in four person offices with quite a bit of room. The other staff members seemed to be very bright, capable individuals. Bell Labs had provided us with a GE 600 series computer which was fast (by standards of the day) and perhaps as good as the best IBM machines out there. Input to the computer was through punched cards, each containing a single line of code. I was provided with trajectory data consisting of range, velocity and elevation data as a function of time for both missiles. I began planning what I would do and proceeded with the work.

While I was at work, my wife began looking for a place that we could rent. Since an apartment was out of the question, she started looking at houses. There were a lot of dirty, run-down places on the market. In the end, it was a choice between a small cottage on Lake Parsippany and a large country estate on seven acres. The estate had been purchased by the county because they were planning to build a dam on the

Whippany River and the house would possibly be underwater. The county was renting it out for the time being. We, of course, chose the estate. It was not all that great, however, because the previous owner had seemingly never cleaned her house (I'm not kidding.). The county made two separate attempts to clean the carpet, but we had to finish the job. The county was actually a pretty good landlord. They were good about repairs and agreeable to any changes we wanted to make.

The house was a modest two-story, three-bedroom in the colonial style, reminiscent of George Washington's Mount Vernon. It was painted white and had four columns in front. It was located in the prestigious Washington Valley a few miles southwest of Morristown. There were many expensive homes close by. Our neighbor on the right was a retired college president. My wife decided to introduce herself to some of the neighbors one day and started with the house just across the street. She introduced herself to the woman who came to the door, who responded with, "I'll see if madam is available." She was talking to the maid. There were no house numbers for any of the houses on our street. Our address was simply "Gaston Road."

The Sentinel project became very controversial when the public realized that the government planned to place nuclear tipped interceptors around every major city in the country. The anti-military and anti-nuclear groups were in a state of near panic and much of the public was distressed by the thought of nuclear weapons in everyone's backyard. This reaction from the populace became a major concern of the government. There were also the SALT negotiations with the Soviet's going on at the same time. In any event, the decision was made to halt development of the Sentinel System and to go back to the drawing board. GRC was ordered to cease work on Sentinel. We continued to be funded indicating that some revised version of the Sentinel System would soon be forthcoming.

I had been working on the Sprint trajectory model and could see that I would need to, in the future, curve fit the trajectory data I had been given. A least-squares curve fit method seemed to be the right approach. I'd heard about this method previously in various contexts although I had no detailed information on how it was done, but I thought, "How hard can this be?" I started with a simple linear fit:

$$y = ax + b$$

The error for each data point is therefore

$$e = y - ax - b$$

The squared error is then

$$e^2 = y^2 + a^2x^2 + b^2 + 2abx - 2axy - 2by$$

The sum of the squared errors across n data points is computed as:

$$Sum(e^2) = Sum(y^2) + a^2Sum(x^2) + 2abSum(x) - 2aSum(xy) - 2bSum(y) + nb^2$$

The next step was to differentiate the summed squared errors with respect to a and then b. Set both equations to zero (i.e., the error minimum) resulting in two equations in two unknowns (a and b). These can then be solved for a and b which are the coefficients of the linear curve that minimizes the summed squared fit error for the n data points. A similar development can be done for a quadratic equation resulting in three equations in three unknowns which can then be solved for a, b and c. Use the same approach for a cubic, etc.

I followed this method and developed formulae for linear, quadratic and cubic curve fits. These were programmed into a FORTRAN subroutine that could calculate least squared error fits for any set of data points. I now had the tool that was needed to fit the Sprint trajectory data.

We did not have long to wait. The DoD came up with a revised missile defense system that was named Safeguard. This approach would defend only Minuteman launch sites and use the Sprint missile. The Spartan missiles were no longer part of the system and there was to be no defense of US cities.

The overall objective of GRC's effort was to develop a simulation of the complete Safeguard system that could be run against a realistic simulated threat, i.e., a large scale attack of Soviet nuclear missiles against Minuteman missile silos. The simulation would perform all the functions that a real system would do in order to identify problems, limitations, oversights, etc. Each member of the staff had a piece of the total system and proceeded to develop the needed models, write the FORTRAN code and check it out on the computer.

Most of the transfers from GRC in SB adjusted well to being in New Jersey and tried to make the most of being taken from the paradise that was Santa Barbara. This was helped by the fact that they were well compensated for their sacrifice. There was one individual, however, (who happened to be in my office) who complained loudly and often about what a terrible place New Jersey was. He often came in to work late or not at all citing a problem with hemorrhoids. It took me a while to remember where I had encountered him in the past. Some years ago, he and his partner had come to our bridge game late. They were loud and may have been intoxicated. They scrambled the cards in one of the hands resulting in a long delay before the game could be gotten back on track. It is management's responsibility to get rid of negative, disruptive workers who distract others and hurt morale. This behavior was allowed to continue for months. I thought they should have sent him home after a week or two.

I was in the habit of smoking a small cigar after lunch. Yes, it was OK to smoke in the office back then. One of the engineers made a show of being highly offended by my cigar and would take his work and retreat to one of the stalls in the men's room as soon as I lit up. What about that smell? Other than that, we were friends and had a regular Saturday golf game. He was also a very competitive person and was constantly competing with me at work and on the golf course. The golf I understood and was OK with that since I usually won. I found the competition at work annoying. He knew little or nothing about my background, and I didn't think he was even in my league.

On the last weekend in January, we went to visit a family we knew back in Massachusetts. It was a pleasant unseasonably warm day as we set out. However, by Sunday, it had turned a lot colder and had started to snow. We decided we had better head for home early on Sunday afternoon. When we finally got on the road, it was snowing heavily and accumulating on the ground. By the time we were near the border with New York State, there were nearly eighteen inches of snow on the road. If the snow continued, there was no way we could have made it home and furthermore we might get stuck at any minute, so we got off the freeway to find a motel for the night and a restaurant. We were lucky as we soon came upon what looked like a suitable place. We rented a room and were referred to the restaurant next door. I asked about parking. The motel owner suggested that we leave our car in the restaurant parking lot for the night. We had a nice meal at the restaurant and walked back to our motel room. The motel room was decorated with pictures of Jesus and plaques citing biblical verses. Hmm! "These are some very religious folks who own this place." The next morning, we were awakened to the sound of a snowplow clearing the motel parking lot. The snow had stopped and the sun was shining. I checked on my car in the restaurant parking lot. It was the only car in the lot and was surrounded by two plus feet of snow. I asked the motel owners if they would plow a path so I could get my car out. They replied, "No way. We hate those people." I said, "Wait a minute. You told me to park there." Nothing I could say made any difference. They weren't going to help me. "What a bunch of effing hypocrites!" I went over to the restaurant and talked with the owners. They were sympathetic to my plight and offered to loan me a snow shovel. Nice people! The car was a good fifty feet from the road, and I set to work. I chained up my car and shoveled, drove forward, shoveled some more, drove forward, etc. until I got to the road. Unfortunately, the highway department snowplows had come through and pushed up a six foot high bank of wet snow and ice along the side of the road. I attacked this obstacle and after another half hour or so got through it. My back was now really hurting. I drove over to the motel and got my wife and son. We would normally drive over the Tappan Zee Bridge to get to New Jersey, but last night during the storm, traffic had been stalled on the bridge and the drivers had abandoned their cars and sought shelter. We heard reports that it would take days to tow away all those vehicles and reopen the bridge. Our only recourse was to drive through the city of New York. We hit the city at about 10AM that morning. After stopping at a few hundred or so lights and getting stuck a few times we made it to New Jersey. Every time we got stuck there seemed to be several citizens there to help push the car. We were impressed and thought, "A lot of nice helpful people live in New York." The roads had all been cleared until we got to our driveway. We had contracted with a nearby service station to plow our driveway when it snowed. They had been out but for some reason had left the last forty feet unplowed. A four foot high snow drift had also accumulated in front of the entrance to the house, but we trudged through it. It was now about 9PM. It had been a long day.

In the evenings after work, I liked to open up a beer and sit down and watch the evening news on TV. First up was the local news out of New York City. The news stations liked to feature stories about the many conflicts going on in the city and always had a number of interviews with really angry people. The network national news was next and although better, still featured some angry voices. After two hours of this, I felt anxious and upset. When I realized what was causing this, I stopped watching the news altogether. This prohibition lasted for years, and we missed out on most of the drama surrounding the Viet Nam War.

We had been in New Jersey for about six months. Everything was going well until July. I came home from work one day and found that it was 80 degrees in the house and not cooling down. My wife and I headed for our local Fed Mart and purchased an air conditioner, which I installed that very evening. We put a screen across the hallway and cooled three rooms allowing us to survive the torrid summer. There was really no lawn surrounding the house (or trees either), but a large area of grass and weeds had been mowed in the past. I purchased a used riding mower from the college president next door and spent a couple of hours each weekend mowing the weeds. In the spring, I had planted a stand of petunias in a narrow flower bed area directly in front of the house. The flowers needed weeding every weekend, and I committed to performing this task until the summer heat hit and the mosquito population exploded. I tried to continue the weeding but had to abandon the effort. In the twenty minutes or so it took to weed the petunias, I would have been eaten alive. The house had shutters on each side of every window, which proved to be an ideal habitat for wasps. With the swarms of wasps, my wife was afraid to let our son play outside.

We had my brother Bruce's Volvo with us and had to register and relicense it in the state. Massachusetts where it was purchased does not issue titles. Your bill of sale is your title. This process involved a lot of writing back and forth with Bruce to get information and signatures. This was stalled when the vehicle failed the DMV inspection. After waiting in a long line, it failed because of a lever on the steering wheel that allowed one to flash the headlights. That was a no-no in Jersey. I went home and thought about it for a while and then got under the dash board, located the offending wire and disconnected it. I think they let me buck the line on the return inspection. The Volvo passed this time. So now, it had a NJ title and license plate. Next on the list was insurance. I soon found out that it is not easy to insure someone else's vehicle especially when he is not even in the US. I decided that it was time to park the car behind the house.

The economy was in recession that summer. The automobile companies were suffering perhaps the most and consequently, were offering major discounts in order to work off their inventories. We needed another car and thought that we could use a station wagon. I researched the market and calculated that a good price for the Ford wagon that we wanted would be about $3600. We visited the dealer and selected the car we wanted with options, which included air conditioning, of course. When we had finished with that task, the salesman said, "How about $3600?" I paused long enough to give the impression that I was thinking it over, giving me time to think about how much less I could get away with. I countered with, "Well, how about $3400?" Afterwards, I was kicking myself thinking I could have gotten a few hundred more off the price. Summer driving with an air conditioned car was a lot more fun. We cruised around the area in our new car enjoying the many historic sites. One such site was in the nearby town of Mendham where we usually did our grocery shopping. The town folks had reconstructed a row of cabins in which George Washington and his troops spent one of those cold winters during the Revolutionary War.

The boss called me in to his office one day and handed me an envelope. It was a notification that I had received an unscheduled pay raise. I tried to ask about the circumstances. Was it because of the good work that I had been doing (apparently)? Did he (or someone else) request it for me? I couldn't get any details out of him. He must have thought it was all too obvious. Anyway, I said thanks. It was a nice gesture.

GRC leased a building in nearby Denville and moved operations to the new location. We had more space and private offices. My work on the Sprint model and computer code was nearly complete. For any intercept point, it could produce an interceptor trajectory that would hit the target, compute launch time and allow interceptor position and velocity vectors to be quickly computed for any time since launch. My continuing activity was to develop interfaces with the rest of the simulation and define communication protocols.

As summer turned to fall, I began to think about the transfer to Santa Barbara that I had negotiated as part of my employment agreement. I called a meeting of the other staff members who would be continuing with my work and working on the tasks that used my product. As I got into my pitch I realized that there were a few paths and connections that I hadn't thought through. I halted my presentation and rescheduled the continuation for a later date. I learned from this that the best way to make sure that you really understand something and identify what you may have missed is to try to explain it to someone else.

One of the vice presidents from Santa Barbara operations was visiting our facility so I scheduled a meeting with him to discuss my transfer to Santa Barbara. He claimed to have no knowledge that such a promise had been made but did not challenge me when I said I had it in writing. He said he would make it happen. In early November, my wife, son and I went to Santa Barbara on a company paid house hunting trip. We found a small three bedroom in the lower San Roque district and rented it for a December 1 move-in date.

The company was managing our move through an SB moving company that was working with the NYC Allied Van Lines franchise. A crew of three guys came to the house and packed our belongings into boxes. The movers, who would pick everything up including our two automobiles, were scheduled for early the next day. The Volvo sedan that belonged to my brother had been put into storage in Newark since he was on another foreign assignment and would return to the East Coast when that was completed. We didn't have an exact time for the moving van to show up, but we were ready early in the morning. By 10AM, no movers had shown up, so we called Allied. They seemed annoyed that we had called and said the movers would be there any minute. By 1PM there were still no movers. We made another call to the Allied people who were now also concerned. Allied said they may have stopped to pick up a ramp for loading our automobiles. Ramps must have been hard to find and acquire. At last, the moving van was coming down our driveway. It was now 4:30PM. The house was about 200 yards from the street. The driveway split as it neared our house with one branch going to the neighbor and the other to us. I went out to meet the van and talked with the driver. He asked, "Does the driveway form a circle?" I said, "No." He then drove down the neighbor's driveway with the wheels on one side of the truck off the blacktop. He stopped and tried to back up. The wheels dug into the soft earth.

There were three men in the moving van, which was painted in the Mayflower moving franchise colors. The driver said that he had recently switched to Allied and would soon get the truck repainted. Hmmm! The three of them got out of the truck and begin stuffing two-by-fours, plywood sheets and even a packing blanket under the wheels. None of this helped the wheels to get any traction; in fact, it seemed to block any movement. I thought about advising them on the basic physics pertaining to getting unstuck but resisted the temptation. I offered to help them locate a wrecker, but they assured me that they could handle the situation. I went into the house to await further developments. After some time had passed, the driver knocked on the door and informed me he was ready for the wrecker. I called around and located a company that assured me they had the biggest wrecker in town. It must have been about 6:30PM when the wrecker showed up. The wrecker driver surveyed the situation and said, "I don't know if I'll be able to get him out." I said, "Could you try?" The wrecker was positioned with its rear wheels on the blacktop driveway and was then hooked on to the van. The wrecker's flood lights lit up the area. A medium-sized crowd had gathered by this time. I thought, "Where did all these people come from?" After a couple of failed tries, the driver gave the winch full power and the van began to move. The van was pulled onto the blacktop and turned around so it was now facing the street. The wrecker's wheels had torn up a section of the blacktop and made some deep gouges in the "lawn." Collateral damage, I suppose. The van driver now attempted to back up closer to the house but had trouble keeping it on the paved driveway and settled for being 40 feet or so from the house. He didn't seem to have a much experience or skill in maneuvering the van.

The next step was to begin loading everything onto the van. It was going to be a long night. The driver wanted to first load our two vehicles, the Mustang and Ford wagon. The movers had only one steel ramp that was about ten feet long and two feet wide. They positioned the ramp with one end in the bed of the truck and the other on the ground. For the other side they built a ramp out of two-by-fours, some plywood and a six foot plank. It looked a bit rickety. The driver tried to drive the Mustang up the ramp but had trouble working the clutch, which was quite stiff and took some time to master. After a couple of attempts, he suggested that I drive it onto the van. Fine! I drove it up the ramp and when the car was about half-way into the van, the ramp collapsed. There was no room to open the car door, so I crawled out the window and jumped to the ground. It was now clear that these clowns did not know what they were doing. I told them to get back as I constructed a ramp under the left rear wheel that was dangling in the air. It put enough upward pressure on the wheel so that it could get some traction. I crawled back into the car and backed it down the ramp and put it in the garage. I told the three stooges that we were done for the night and I would call Allied about what to do next. Well, maybe it was the two stooges. The third man was a local helper who seemed the most rational. My apologies to him.

When I talked to the Allied dispatcher the next morning, I told him I wanted someone else because the crew he sent out was a disaster. He said there was no one else, and I was stuck with them. We had to cancel our flight reservation for the day. The new tenant was also scheduled to move in that day. Allied did come up with a revised plan where our two cars would be driven into NYC and loaded from the company dock. The movers would rent a pickup truck and shuttle our belongings from the house to the van parked on the street. Fortunately, we had not yet accumulated a lot of furniture. This was a big improvement over what happened last night, so we went along with the new plan. The operation went forward with no further problems although it seemed to take a long time. The new tenants waited patiently with their moving van (a smaller and more maneuverable one) parked on Gaston Road. By late afternoon, the loading was completed and we drove into Morristown to our hotel and said farewell to the movers as they headed for NYC with our load of possessions and two automobiles. We were greatly relieved as we sat down to dinner in the hotel dining room. The whole affair had been very stressful, especially for my wife who was eight months pregnant and my young son who was distressed by having his life uprooted and seeing strangers moving into his house.

Looking back, our three year East Coast swing had been quite an adventure. We had visited seven states for the first time and had seen many historic sites. We experienced the East Coast cultures of Massachusetts and New Jersey. I had gained expertise in several new areas including: strap-down inertial navigation, reentry vehicle penetration aids and missile defense. My salary had been increased by over 50%, and we were saving money. We might have stayed longer, but the opportunity to move to Santa Barbara was too good to pass up.

Santa Barbara

On Thanksgiving Day in 1969, we flew to Los Angeles and rented a car for the drive to Santa Barbara. GRC put us up in a good motel until our furniture arrived and we could move into our rental. We contacted the local moving company that was handling our relocation. They had encountered a bit of a problem. They couldn't find the movers who had our stuff. It was several days later that they found them in San Diego. The movers claimed to have stopped at "home" – not exactly on the way. They eventually showed up and moved our belongings into the house we had rented. The only damage to the furniture was a long scratch on the surface of our mahogany dining room table. The local mover's representative said that there was a local firm that could put a veneer on the surface to fix the problem. Unfortunately, both cars were damaged. The tops had creases along both sides indicating that something heavy had been placed on top of them. The local moving rep. again had a solution, which was to take them to a local body shop to have the damage fixed and both cars completely repainted. He said that our moving crew needed to learn a lesson to take better care of people's possessions. He said that they would also be billed for the damage to the blacktop and the yard of the house we had moved from. He was showing these guys no mercy. I think he was highly pissed over the whole affair. I don't think our movers made any money on this trip. Looking back, I think they were running some kind of "free-lance" operation, that is, they were picking up and delivering for customers that their employer didn't know about. They could also have been running contraband, maybe drugs. Clearly, they were doing something off the books, which would explain why they were always late.

The house was a modest three bedroom in a neighborhood of similar homes. It was built just after WWII by my estimate and was on a small lot. On the outside, it had been painted pink with light green trim and that color scheme had also been used inside with green carpet and pink walls. It was perhaps my least favorite color combination. We had only a one year lease and didn't plan to stay there any longer than that, so it was good enough.

Although the baby was not due until mid-January, we decided to forego the New Year's Eve party being hosted by one of the engineers we had known in New Jersey. It was a wise decision since Carole went into labor in the early morning hours. We rushed to the Santa Barbara Cottage Hospital and our doctor was called in. Our daughter, Cheryl, was born at 9:30AM and was the first baby born in Santa Barbara in 1970 and the second in the county. The first baby born in the county received lots of prizes – the second not so much. Our doctor was an alumnus of the University of Michigan and had tickets to the Rose Bowl football game. He joked about missing the game but delivered our baby and still had enough time to drive to Pasadena. Unfortunately for him, USC proved to be the better team by the score of 10 to 3.

I checked in at the Santa Barbara GRC facility, which was actually in the nearby town of Goleta. They gave me a nice office with a window. There was a secretary I could use sitting just outside my office. I had brought about six months of work for the New Jersey project with me and thought that I was all set, at least, for a while. After a couple of months, I realized that I was having no contact with anyone in Santa Barbara operations and began to wonder. "Why were they ignoring me so completely?" I asked for a meeting with the head of operations and went to see him. I introduced myself and he said, "Who are you?" Great! He doesn't know anything about me. I went through a list of my skills and relevant experience that I thought might give him an idea of where I might fit into the organization. He was visibly unimpressed. He said, "Well, we are going to lay-off some people this summer. You might be one of them." I tried to make some sense of the situation but came up empty. Let's see. The company had spent a lot of money bringing me to Santa Barbara – a house hunting trip, moving the family and all our possessions to SB, putting us up in a motel for ten days, etc. I could have gone to the VP who had approved my move to try to get some answers, or I could contact some of the people who had been in New Jersey and had returned to SB. I chose the latter and went to talk with Roy who had been the lead on designing the simulation in New Jersey and was something of an assistant manager. He talked with Larry, the manager of the Mathematics Department, who agreed to give me a trial assignment. The group was working on a simulation called the Ballistic Attack Game (BAG), which looked at attacking forces of ballistic missiles and defense strategies. The current version, BAG 14, was in development and needed an interceptor missile component. This was a lot like the Safeguard system I had worked on in New Jersey and was right up my alley. I began adding the interceptor capability. The procedure was to submit one's computer run to the data center by the end of the day and pick up the results in the morning. When I came into work in the morning, I could find neither my run output nor my deck of cards. The data center didn't have my stuff, and no one seemed to have a clue as to what had happened to my run. Then it would magically appear later in the morning on the table in front of our secretary's desk. I finally figured out what was happening. Larry had been intercepting my program and checking out what I was doing but never saying anything to me. I found this rude and disrespectful behavior highly annoying. Apparently, I passed muster because this BS stopped after a few days. I was transferred into the department and began officially working on BAG 14.

One of the staff, Pete, had a contract to be part of a study team which was evaluating a missile defense system that was much like the Safeguard System. This may have been a DARPA project. He was shuttling back and forth to Washington, DC to attend meetings. He had agreed to provide performance results for the system using our simulation but had neglected to ask for help from the Mathematics Department to produce those results until a week or two before the deadline. We went into panic mode to get something up and running quickly. Bag 14 was an event based simulation that used dynamic storage in which a function asked for storage in which to save data and then released it back to a central pool when no longer needed. This approach had been developed by Roy and made communication by different parts of the code much less problematic. For the event based aspect, the simulation works off a calendar of events arranged in time sequence and executes the next event on the calendar. Any function in the program can schedule an event and put it on the calendar.

Our objective was to be able to handle a string of about 30 attacking RV's by intercepting them with Sprint missiles that were armed with nuclear warheads. The RV's were spaced a few seconds apart so that the defense would not be able to kill more than one with a single interceptor. Larry was in charge of target acquisition and radar tracking. Once a target was put into track, he would create an intercept planning event. That's where I came in. I received no guidance (pun intended) as to how any of this was to be done, and it took me a while to realize what the scope of my job was. It was what to do and how to do it so that the attackers were killed. I now realized that they were counting on me to be an expert in engagement planning and missile defense strategy or were hoping I was that guy. I began with the most basic functions such as: define missile battery locations, engage all of the missiles in the string, compute intercept points for each, etc. I got no input from Pete as to what the community consensus on how to do these things might be, but I pressed on. I decided to engage the first missile in the string at the lowest intercept altitude I would use (e.g., 50 kft.) and the nth missile at the highest (e.g., 300kft.). With an altitude spacing of 50 kft., I could intercept 6 RV's at a time. This allowed time for the nuclear effects to dissipate so that the radar could track the next batch of attackers as they came in. This approach would keep recycling the intercept points until all attackers had been engaged. I called this approach "ladder down," and did not find out until much later that this was one of the approaches being used by other companies who were looking at the problem. The terms "ladder down," as well as, "ladder up" were being used. Looking back, this is a simpler problem than it seemed at the time. I think a paper and pencil analysis could have been done to develop a strategy and predict results. But we had chosen to do it with a simulation which meant we had to go through most of the functions and steps that a real system would have to do. Doing it this way forces one to really understand the problem so it was probably a good thing. We labored on and as the deadline approached, we still had much to do. I was "asked" to come in on Saturday to work on the problem. Larry and I labored all day Saturday and on into the night writing new code, fixing bugs and running the program. It wasn't until

Sunday afternoon that we got our first successful run in which we acquired, tracked and destroyed a large number of the 30 RV's in the attack. We had worked thirty hours straight. It took a few more days before we could generate some results that Pete could use. He rushed back to DC to share them with the study group. The group seemed to appreciate the information, but he got marked down for being late with his results. The lesson here is to not wait until the last minute to start work or ask for help.

Santa Barbara is one of the most desirable locations on the earth to live due in large part to the excellent climate. Winters are mild so that outdoor activities can be enjoyed year round. There is some rain in the winter months, but most days are mild and sunny. Summers are warm but not hot thanks to the cloud bank that rolls in at night and doesn't dissipate until early afternoon. The city is quite isolated from the LA area and not subject to the congestion that clogs the freeways there. Santa Barbara was enjoying a building boom at the time. There were several new housing tracts being built west of the city in the Goleta suburb. We had a one year lease on our rental and had no intention of renewing. Carole shopped the new developments and decided that the Mountain View tract had the best style and amenities as well as the best location, being bounded by Cathedral Oaks Road and Kellogg Ave. We purchased a one story four bedroom with about 2000 square feet for $35000. That house recently sold for over one million dollars.

The yard was not landscaped and I knew little about what to do, but since one does one thing at a time, there was time to research everything. My first project was to plant the front and back lawns, which I, of course, did on weekends. I rototilled, added soil amendments, leveled, planted grass seed and watered. No one could have anticipated all the weeds that grew up among the new blades of grass. I spent many hours sitting on a sheet of plywood on the lawn picking out each and every weed. As the grass grew, there were large bare areas. A neighbor advised me that what I needed was Scott's Turf Builder, so I hand cast this product all over the lawn, and "bingo" the bare spots began to fill in and the whole lawn began to look healthy. There were many different ground covers that are used in the area; ajuga, gazanias, freeway daisies, ice plant, etc. I tried all of these for the border and bank areas and settled on freeway daisies and ice plant as the hardiest and most trouble free. Once ice plant gets established, nothing can kill it. I tried to get by without lawn sprinklers but eventually gave in. I had built a 10 foot long pipe with 3 sprinkler heads on it with the intention of moving it around as needed. One of the neighbors saw my wife dragging this apparatus across the lawn one day and told her I should be ashamed. Point made. I took 2 days of vacation and dug trenches in the front lawn and installed the pipes and sprinkler heads. The lawn liked the new system and I had the greenest most luxurious lawn on our block. I installed sprinkler lines and rain birds in the back lawn which was the play area for our kids.

The BAG 14 simulation provided a platform for studying various ballistic missile phenomena. RV and penetration aids wake characteristics had become the focus of several DoD studies. Our Huntsville Division had a contract with the Army to model radar returns from various combinations of RVs and penetration aids. We modified BAG 14 to include this capability and one in which the radar ran in burst mode. By passing the returns data through a fast Fourier transform program, they hoped to separate the incoming objects in range and velocity. We had obtained an FFT (fast Fourier transform) program from our division in New Jersey and were to make it available to Huntsville. There was a female programmer, Jenny, in our group who was assigned to check out the FFT program and make sure it was working properly. The program had been delivered to us with a printout of a sample run but no documentation. Jenny got the program running on our computer and was able to run it to reproduce the output for the test case. "I was able to match the test case," she observed. "It looks like it is working." Roy was skeptical and asked me to look into it. I looked at the test case, which was page after page of meaningless floating point numbers telling me nothing. I decided to run a simple test. I input some data for a sine wave expecting to get a couple of spikes at the wave frequencies. It failed. All I had was a FORTRAN listing so I went through it. It was something of a guess, but I began to suspect that there was an error in an IF statement defining a branching decision. After fixing the supposed error, it worked. I tried a few more simple waveforms and got consistently correct results. I told Roy it was now ready for prime time. Jenny had failed to do her job. She also had a habit of coming in to work late – claiming she was working at home. She was gone in the next layoff. Using the modified BAG 14 and the FFT program, our Huntsville people put together an impressive color slide show presentation showing simulated radar returns for different kinds of reentry objects plotted in range and velocity.

Some people in the company liked to brag about the superior work ethic of the staff – long hours and extra effort. I didn't see much of this. Perhaps that was something from the past when the company was first getting established. I made it a point to come in at 8AM or a little before – the scheduled start of the work day. I was always one of the first to show up. I didn't leave until nearly 5PM. Nearly everyone had cleared out by that time. The work force was not putting in their hours and, therefore, not being very productive. This made our services more expensive and was likely to be a problem down the road.

There was one manager who had a reputation for putting in long hours and because of this was thought to be something of a hero. I had to work late one evening and was able to observe him. He spent a lot of time in the coffee room and wandering around. He didn't seem to be doing anything that could be called work; in fact, I got the impression he was just hanging out at the office. Maybe there was trouble at home and he needed to get away from the conflict.

Another use for BAG 14 was to model certain nuclear effects. A nuclear phenomenology and radar propagation model called RANC IV was added. The model was based on measurements made during the US nuclear test program and were based on real observed effects. Scientists at Lawrence Livermore Laboratory took an interest in the code and gave me a contract to make changes to the program and run various engagements. We exchanged visits, and I made several trips to the Livermore site. RANC IV was an Atomic Energy Commission (AEC) sponsored code and there was a document that contained some recent updates to the model. My boss decided that I needed an AEC "Q" clearance so that I could review this material. This was a high level clearance on a par with a DoD top secret clearance. I filled out the proper paper work, which touched off a rather thorough investigation that included interviewing my neighbors. My next door neighbor was Polish and had been in one of Hitler's work camps. He even had the tattoo. I'm sure his pulse quickened a bit when the Federal agents knocked on his door. My clearance was granted in time. Protocol required that a closable curtain be installed on my office window and a deadbolt on my office door. When I wished to view the document, I had to go to our security department which was the custodian of the document, procure it and return to my office. I would then bolt the door and pull the curtain before reading it. There was a minor stumble along the way. There was a young woman employee whose first name sounded a lot like my first name initial plus my last name. She was a bit flirty and considered to be sexy. The maintenance worker made the installation in her office first. Perhaps he was just hoping it was her and not me.

My brother, Bruce, had given me a stock tip about a company called Cardiodynamics, which was developing a heart monitoring device that a patient with heart problems wore on his chest and would alert the wearer if there were an irregularity in heart rhythm. This was a revolutionary concept. The stock got a lot of notice and began to take off. I bought a few hundred shares at $2 each. The company was located in the SF bay area and located between Livermore and the airport. On one of my trips to Livermore, I stopped in to see the operation. They had two rented rooms in a strip mall, and the only person there was the secretary. She explained that the two founders were professors at UC Berkeley and were at work. They came by on weekends and worked on prototype devices or traveled around the country pitching the device. It was hard to see much value here. I sold my shares as soon as I got home and made a good profit. The company soon failed and many investors lost everything.

The Sentinel ICBM defense system had been canceled in 1969, but interest remained high in finding an effective way to defend US cities from an all-out missile attack from the Soviet Union. A study group was formed in the fall of 1970 to look at options that would address this issue. The Boeing Company in Seattle hosted the study group comprised of several companies to identify and evaluate various options. This activity was given the name of "The Indian Summer Study." GRC was one of the participating companies and was tasked with evaluating the effectiveness of the different approaches that the group might define. We had a computer program that evaluated defense effectiveness against specified attacks. My role was to modify the program to model the attacks and defense systems defined by the group and run the different cases. I made two trips to Seattle in support of the study. The GRC team was allocated some space in the headquarters building at Boeing field. Boeing was suffering from a major downturn at this time and had laid-off a large number of workers. During one of my lunch breaks, I made a self-guided tour of the building. Each of the upper floors consisted primarily of a large open bay. No one was working there. All the desks and chairs had been pushed into a giant stack in the center of the room. Depressing! The study team defined a preferred system that divided the country into sectors and allocated a number of defensive missiles to each sector according to the population it contained. One of my computer program inputs was a large card deck with "collateral damage" numbers for each and every population center in the country. Our final system design proved to be effective and limited the collateral damage estimate for the whole country to one half million. I thought this result a bit strange since one hit would take out a million people. The answer was that the program determined that the probability of even one attacker getting through was only 0.5. What it was really saying was that it was equally likely that the collateral damage was zero or one million.

On Fridays at the end of the working day, there was usually a company sponsored party in the auditorium with free beer and snacks. I nearly always avoided these social gatherings preferring to go home to my wife and kids at the end of the week. Drinking with one's co-workers is inherently danger-ous since people may behave badly under the influence of alcohol and it was hard to see how anything positive could come of it. The party on the last work day before Christmas had a tendency to really get out of hand. There was a man-ager of one of the groups who had been with us on the Indian Summer Study and who was very likeable and seemed to have a bright future with the company. He had too much to drink at the Christmas bash. He went to the lobby and called his wife to come and get him. Too late! He threw-up all over the carpet. The company's treatment of this kind of misstep was usually harsh. He was gone within a few days. There was probably more to the story.

There seemed to be someone in the company who was in charge of morality enforcement and who set a fairly high stan-dard. At the company picnic one year, a young woman, who was a tech support person, came to the event braless. This was at a time when this look had gained some popularity with the college girls in the area and other parts of the country. She gained a lot of notice and received some attention from many of the male attendees. She must have offended the morality enforcer, and she was gone the next week. Some members of the salaried technical staff engaged in questionable behavior from time to time, but they seemed to be mostly immune from any company action.

Most companies do not allow alcohol on the premises, at least, the larger ones. GRC was different and we liked to celebrate the holidays with Saint Patrick's Day being one of our favorites. We took up a collection and bought a fifth of Irish whiskey. There would usually be some left over which I would save in my file cabinet for some future occasion when everyone was down or there was something to celebrate. The company sponsored an annual Christmas party and furnished all the food and drinks. The preferred venue was El Paseo, an historic Mexican restaurant in downtown Santa Barbara. Everyone seemed to enjoy this in principle, but there was always a band that played so loud that it was painful. We were forced to retreat to the farthest reaches of the restaurant in order to have any conversation.

When I joined Santa Barbara Operations, there were about 150 members of the technical staff, our official title. Every summer when the company balanced its books for the fiscal year, a company-wide meeting would be called to discuss the state of the company finances. This was usually accompanied by an announcement of a lay-off of two or three people. The reason given was that the lay-off was necessary to balance the books. This didn't make much sense. A well-run company should lay-off the poor performers every year. Was this what they were doing but were afraid to come out and say so? The company was shrinking in size every year. Just before Christmas one year, the Army program that funded our missile defense work got cut by 30% resulting in an identical cut to our funding. This was a large part of our budget, and the company reacted by laying-off 30% of those working on the project. The company tried to postpone the news until after the Christmas party, but one of the guards who had seen the lay-off list started talking at the party. This ruined the celebration for some.

When someone in the Mathematics Department was laid-off, their card file was assigned to one of the remaining staff members. A card file is a large steel cabinet that contains the history of a programmers work experience in the form of punched cards. I inherited a number of these over the years. In general, I found that they were for the most part useless as I seldom had the occasion to consult any of them.

An Army major was our contract manager and liked to come to Santa Barbara and play golf as part of his inspection duties. It was GRC's task to put together a foursome and schedule the event. I was once asked to join the group and, of course, accepted since who can turn down free golf during working hours? We played at Sandpiper, a seaside course and one of the finest private clubs in the area. We were all average golfers (able to break 100 on a good day) with the exception of Joe, who was the Montecito CC champion. My game was erratic until we hit the back nine and things started to click. I was driving with a three wood and started hitting it long and straight. Joe was outdriving me by a bit but was hitting his driver off the tee. As we came to the 18th tee, Joe and I were tied on the back nine. It was a 155 yard par 3 with a large pond in front of the green and a large sand trap guarding the right side. I hit a 5 iron onto the green about 25 feet left of the hole. When Joe put his tee shot in the trap, I thought I had him. I was able to 2-putt, but Joe blasted out of the trap and sank a long putt. Well, he was a club champion. I was pretty happy with matching his score for the back nine. The major didn't play very well that day but seemed to enjoy the outing.

The Mathematics Department took on an ambitious project called the Adaptive Tester. This computer program would allow users to test missile defense software at the module and complete system levels. My role in this was to construct and document the Automated Scenario Generator. The ASG accepted a menu of attacking missile threat descriptors and targets from which it generated a set of trajectories. This was an easy task for me since I had many of the pieces in my personal library of programs from past efforts in this area. I concentrated on turning out a quality product.

A staff member who I knew fairly well got axed in the summer lay-off that year. I had known him back in New Jersey and we were on friendly terms, often talking about work issues and the passing scene. Our wives were also friends and part of the GRC wives social group. I don't know what his performance shortcomings might have been; however, he did have a meltdown back in Jersey where he launched a tirade in the computer room one day, yelling incoherently at no one in particular. It was a self-service setup, and someone may have bumped him in the queue, but still it was an extreme reaction. He found another job at a small firm in LA that did (or hoped to do) some missile trajectory modeling work. He contacted me from his new job and asked for my help in putting together a trajectory simulation computer program written in FORTRAN. I was concerned about his request from a conflict of interest point of view but decided to help him anyway because he was a friend. He offered to pay me $200 for my help. I would have helped him for nothing. He came over to the house on Saturday of the Memorial Day weekend and we sat down to write some code. I had just finished such a program for the ASG so it was fresh in my mind. I wrote from memory and spent 4 or 5 hours putting together a program which would do exactly what he said he wanted. It was in many ways identical to the one I was running on the GRC computer. A few days later, I received a phone call from him informing me that they couldn't get the program to run and it was probably because what I had given them was no good. I then spoke to the "boss" who also informed me that my work product was of poor quality. He specifically mentioned the Runge-Kutta integration routine claiming it was bad code. Also, they wouldn't be paying me anything. This was all quite strange, especially since the Runge-Kutta routine, actually a clever piece of code, was from the company library and something we all used. If they had asked for my help instead of slamming my efforts, I would have gladly gone down there and probably gotten the program running in an hour. Now I really didn't know what to think. I concluded that they were either incompetent boobs (probably), or they were giving me this ridiculous story as an excuse to not pay me (possibly) and decided to cut my loses. I never heard from my "friend" again.

The building boom continued in the Santa Barbara area but was running into trouble. The water source for the area was the Lake Cachuma reservoir in the mountains just north of the city. Santa Barbara County water consumption was increasing due the demands of new construction and the supply available from the reservoir would soon be inadequate to meet the county's needs. The county water board was populated with local ranchers who were making big profits by selling their orange and avocado groves to developers. Their long term solution was to connect to the California State Water Project; however, this would require an aqueduct to be built to carry the water over the mountains from the central valley at considerable expense. Local activists did not like the future being created by this scenario that would result in a much larger population accompanied by more congestion and reduced air quality. Four young professionals collaborated to get on the ballot and throw out the rancher coalition at the next election. My wife and I supported their efforts, attending a number of meetings and contributing to their campaign. One of the candidates was a GRC staff member, a young civic-minded PhD. The reformers were successful and now dominated the water board. They stopped most of the new development. There are often unintended consequences to seemingly responsible acts by government. In this case, housing prices began to rise substantially due to the building moratorium and were also pushed higher by the 10 to 15% inflation in the economy.

My son, David, was in the third grade at the Mountain View School that was just a block and a half from our house. It was a fine school that featured open classrooms and a progressive approach to education. The school decided to have the third graders perform calisthenics during recess, doing a variety of movements. David complained that doing one of these, the so-called four-count up and down (aka squat-thrust), caused him to have hip pain. The calisthenics instructor responded by having him sit out during this one move. This was OK except that he was made to sit in front of the class while they performed the exercise. I heard about this and thought that it seemed like this treatment was designed to embarrass him and punish him for complaining. I went to the school and talked with the principal. We had a lengthy talk, however, I got the feeling that he was just humoring me and intended to do nothing. I decided to follow up our meeting with a letter. A letter requires a written response to the writer and must be passed up the line. I explained in my letter that the callisthenic movement in question was part of the Army PT (physical training) test given annually to soldiers to measure their fitness levels. It is designed to be a stressful exercise. I also stated that calisthenics were probably not appropriate for third graders. Within a week or two, the callisthenic program was canceled.

Goleta was home to a number of high-tech industries working on defense contracts. Many of the scientists, engineers and technical experts who worked at these companies chose to live in the Mountain View neighborhood. Many of these young professionals had small children who attended the same school as my son. The school sponsored (probably a state requirement) a "gifted" student program. Starting in the third grade, all the children were given an intelligence test. Those who scored at the near genius level were placed in the gifted program. It was a point of pride for parents to have a child placed in the program and, conversely, a major disappointment and even embarrassment if one's child didn't make the cut. The percentage of children who qualified for this program was quite large for our school compared to the general population and reflected the makeup of the neighborhood, populated by many young professionals. Our son passed the test and participated in the program.

The real estate market in the county was booming and prices were going up rapidly. I became interested in real estate, thinking it might be a good area for investment. However, I didn't feel confident buying anything until I knew more about the subject. I enrolled at Santa Barbara City College and began taking real estate courses, and eventually took the eight or so courses offered and earned a certificate in real estate. I decided to take the real estate salesman exam. There was a company in the state that sold lists of questions that had appeared on past exams. I bought the list and began going through the hundreds of questions. The first time through I got about 70% correct, which is the percentage one must get to pass the exam. On my second pass through, I hit 90%. I went through it a third time, for good measure and got close to all of them right. I was ready to take the exam.

In the summer of 1975, we went on a family vacation to the San Francisco area. We spent a day in the city riding the cable cars and seeing the sites. We next headed to the Napa Valley and spent a few days there. We stopped at some of the wineries and others points of interest. Even though it was the tourist season, the valley highways were uncrowded and wine tasting was free at all of the wineries. We drove around Napa to see the town and do some house shopping. Napa had a small town feel and real estate prices were slightly lower than Santa Barbara. It seemed like a nice place to live.

At GRC, I continued my work on the ASG and also on a more detailed model of the Sprint interceptor. I added the capability of maneuvering in response to guidance commands. The

Sprint missile would be tracked by the phased-array radar after launch and be sent commands to turn the interceptor and keep it on a course that would intercept the target. We were requested by our US Army customer to share this model with other companies. We got a conference call one day from one of these companies requesting information on the sign convention of the turn commands. It was whether a plus command would turn the interceptor to the right or the left and also up or down. I tried to explain the rule, but it seemed that it was just not getting through. Finally, I suggested that they put the model into their simulation and give it a plus (or minus) command and see what happened. That seemed to satisfy them.

In my last performance review, there was a comment by my boss that said, "Julian thinks that he is more than just a programmer." I was stunned. I should have asked, "Just what the hell do you mean by that?" but I didn't. I was as he said "programming," but I had always created the content. I had also reviewed and fixed various modelling errors and omissions by others in the department. To be honest, I had dreaded and knew that this day would eventually come, the day when I would officially be tarred as a programmer and not be an engineer. At the two companies I had worked for previously, I had been the top rated engineer in my group. Programmers are lower on the pecking order than engineers. I had accepted the situation at GRC because I was being well paid and the job allowed us to live in beautiful Santa Barbara. Another problem with my job was that my pay had effectively been cut, especially in the last three years. The Federal government had placed a cap of 5% on salary increases in the aerospace and defense industry. Inflation was running at 10 to 15%. I calculated that I had lost 25% percent of my salary because of this. There were 150 members of the technical staff when I joined Santa Barbara operations six years ago. This number had shrunk to about 100. Was it because the company was being so poorly managed that it was no longer growing and was actually shrinking? I began to think that I was in the wrong occupation and certainly with the wrong company.

After 16 years working as an engineer, I had never really bought into the idea that I would do this forever. I didn't really fit the profile of an engineer anyway. I had an aptitude for many things and probably could have been successful in many other professions. I was an engineer because that was my first job out of college. I had broad interests that included real estate, the stock market and current events. I read the local newspaper and the Wall Street Journal. After I had been working for about 3 years, I took the US Foreign Service Officer test given by the US government. I passed the first test and was invited to take the second. If I had pursued this, I could have been hired by the Foreign Service or maybe the CIA. I would have had a much different life. I seriously considered this option but in the end, declined to follow up. Of course, I might have found myself working in the embassy in Tehran in 1979 – 1980. The Foreign Service is actually a very safe occupation although the CIA is not so much.

Men are known to go through what is called the "mid-life crisis" that seems to occur around age 40, my current age. The most likely response for an individual in this circumstance is to change careers to something totally different, to change jobs within one's current profession or to make the focus of one's life to be an activity outside of work. I decided that a complete change of careers was in order. I thank my wife for being very supportive of this decision and going along with it. My plan was that we would sell our house in Goleta, move to Napa and I would try to sell real estate. My backup plan was to find work in the wine industry so that we could eventually start our own winery. Financially we were in pretty good shape, having saved and invested in the stock market. The real estate market in Santa Barbara was booming and selling our house would net us over $50K. In all, we had five or more years of income. Even if none of these options worked out, I could always return to the aerospace industry.

I turned in my resignation to the head of GRC Santa Barbara operations. He paid me a visit and seemed to think I must be kidding. I had mentioned a 25% raise (to recover what I had lost due to inflation). He seemed to know who I was this time. I got the usual farewell party that was held in one of the conference rooms. It was well attended and a very nice sendoff. I was surprised that so many people in the company liked me (or at least didn't dislike me) enough to attend. My new boss, who had taken over the group in a recent reorganization, said that if I ever wanted a job I should call him.

The company had just awarded annual raises. I was given the usual 5%, but it didn't show up in my last paycheck as it should have. I contacted human resources and asked, "What happened to my raise?" Their response, "Well, since you're leaving, you don't need it." I sensed a little retaliation there and said, "I'm still working every day, and I'm going to finish my project before I leave." They restored my raise.

I was working on the ASG module that was mostly my work

and responsibility. I had finished the code earlier. It was working so I proceeded to put my final touches on the documentation and produce a preliminary draft. The document would be finished when the other pieces that were to be created were done and ready to integrate into the final version. A programmer was assigned to take over my ASG work. I don't think my manager appreciated the complexity of my trajectory generation program or realized that some expertise in ballistic missile trajectory modeling might be crucial. The chosen programmer had no experience at all in this field and was going to have a lot of difficulty understanding what I had done, but that was no longer my problem.

We decided to sell the house ourselves without using a real estate agent. With my real estate coursework, I felt confident that I knew what to do. I put an ad in the local paper and put up yard signs advertising our open house. We drew large crowds and received a full price offer during the open house. Well, we should have asked for more. We accepted the offer, and I started an escrow. There is one advantage (I found out later) to using a real estate agent, that is, he/she acts as a buffer between you and the buyer. Our buyer kept coming around asking for more stuff. He even wanted me to replace the shower because of a squeak in the floor. I finally had had enough. I told him I would give him back his deposit, and he could find something he liked better. The harassment then stopped. We hired the local moving company that had moved us to Santa Barbara. They loaded up our household items and furniture and said, "See you in Napa."

Napa Valley

CAROLE AND I went on a house hunting trip to Napa a few weeks before I left GRC. We looked at both new and older homes and decided to go with the new and bought into a tract of homes under construction. The house was equivalent to what we were moving from – a one story with four bedrooms. We added air conditioning since we knew there would be some very hot days in Napa in the summertime. The house was on a corner lot which adds value in the Napa area since it provides easy access to the back yard for storage of a boat or trailer. The distance between the house and the side property line is often tight making it hard to access the back yard for those not on a corner lot.

After moving in, the next thing to do was to begin landscaping the yard. I rented a rototiller and began tilling the front yard. I soon realized that I was going to need a rototiller several more times, and the cost of renting would be as much or more than if I bought my own machine, so I went to the garden supply store and bought a 5 HP model, which I thought would be adequate for my needs. This time I knew to install sprinklers before putting in the lawn. I had also learned another lesson with the Santa Barbara house and decided to buy turf instead of planting grass seeds. My wife and I spent a few hours one afternoon unrolling sod and "voila," instant lawn. We were the first in the tract to have a lawn. It was a magnet for the neighborhood kids until the rest of the home owners caught up. It was traditional in tracts such as ours to put in a fence on each side of the house and in the back of the yard. The cost is typically shared with each neighbor; at least, that was the idea. In Santa Barbara, the majority of the wives were "stay at home moms." In this development, most of the families had two earners, and they were probably less affluent and less educated. My neighbor on one side refused to share in the cost of the boards. He was OK with the posts and stringers. His reasoning was that he wanted grape stakes instead of redwood boards. I took what I could get and went ahead with the fence. In the time that we owned the property, he never put up his grape stakes. There was another problem with the same neighbor. Each lot is supposed to be graded so the runoff during the rainy season will drain to the street. Every time it rained, a pond would form in my backyard from the runoff coming from his yard. I thought about negotiating this issue with him, but considering the response I had gotten on building the fence, I decided to fix the problem myself and built a dam along the property line that would funnel his runoff into my drainage ditch and to the street. This worked well. I don't know if he was even aware of what I had done. Lesson learned: when dealing with uncooperative people, it is often easier to just work around them. I made a patio cover out of 2x4's and corrugated roofing panels. The patio area was landscaped with jasmine and a mulberry tree. Also, a few shrubs and trees were added to the yard. The remainder of the landscaping would be done as time permitted over the next several months.

David and Cheryl were enrolled in the Northwood Elementary School that was less than one half mile away. David was in the fourth grade and Cheryl in second. The children were able to ride their bicycles to school over the lightly traveled streets in the development.

I had spent the first two months in Napa fixing up the new home and was now ready to begin my career in real estate. I chose a small franchise to check out first. It was a family (father and son) business with a dozen or so agents. After a brief interview, I was "hired." Real estate agents are paid by commission, which typically is 6% of the sale price. This amount is divided four ways. The listing agent and selling agent get a share and the broker/owners of the firms representing the buyer and the seller each get a share. The 6% commission may seem like a lot, but the selling agent is typically only going to get 1.5% - not that much. I jumped right in and began pulling floor duty once a week. The person on the floor is supposed to get the "walk-ins." A lot of people seemed to think they had the right to waste the salesman's time and have him/her drive them all over town looking at various properties even though they were not serious buyers. They found the tour to be an interesting way to kill a few hours or even a whole day. I got quite a few of these but didn't spend much time qualifying and screening them because I wanted to get experience working with people. The real estate market could be described as "hot" at the time and a lot of properties were being bought and sold. This was good for sales people, but it attracted large numbers of them to join the business. The field was crowded. It seemed like everybody and his brother were trying to sell real estate in Napa.

Other than working with the walk-ins, the way to get listings or sell property is to cultivate friends, neighbors and just about anyone who crosses your path. There was a house on the east side of town just a block or so off the Silverado Trail that I thought was the least desirable house in town. It was small with uneven siding and hadn't been painted for decades, but people were living in it and it was someone's rental property. Well, it turned out it was owned by my neighbor. There was a "farm" that touched the right-rear corner of our lot. This property contained about 5 acres and was owned by an elderly lady. The farm had been in the family for decades and she had no interest in selling. She employed a caretaker who lived on site with his wife and two young girls. We got to know this family as their daughters often came over to play with Cheryl. We thought of them as less sophisticated but well-meaning and friendly. The guy was actually a bright fellow who saved his money and bought houses to rent out. He was the owner of the ramshackle house. We talked real estate and he expressed interest in upgrading his holdings. I think he didn't trust real estate people and had never gotten to know any of them. Now he knew me. He also owned another rental in a somewhat better part of town but still one of the least expensive homes in the city.

He gave me my first listing. As I excitedly rushed to the office to file the paperwork and make it official, I got a speeding ticket – going over 35 mph in a 25 mph zone. This was a shock because I had never seen a speed limit sign on Trower Ave. even though I frequently traversed it. I went back to where I had gotten the ticket and looked. There was a speed limit sign but the view of it was obstructed by a telephone pole. I took a picture. When I went to court, I told the judge that I had taken a picture of the sign but that it was hard to see. He looked at my picture and said, "Where is the sign?" I replied, "It's behind the telephone pole." "Case dismissed." My victory was sweet. Lesson learned: check out all of your options before giving up.

I eventually listed both of my neighbors' rentals and later helped them reinvest the proceeds in a better rental. This property had been on the market for some time, probably because of its location – on a very busy and noisy street. The listing was held by one of the agents in my office. Our owner/ broker asked everyone in the office to inspect the home and estimate what it should sell for. My estimate was the lowest and seemed to upset the listing agent. The irony of the situation is that I got my neighbor to make an offer on the house at my estimated price – the only offer that was ever made. The listing agent and our broker put a lot of pressure on me to get my client to increase the offer, but I held firm and it sold at the offer price. A house is only worth what some buyer is willing to pay for it. The listing agent never forgave me.

My real estate career wasn't exactly taking off although I did have a few sales and listings. The broker who owns an office can make a lot of money if he/she has a lot of sales people out there beating the bushes since he gets a cut of every deal they can make. I could open my own office if I had two years of experience as a salesman and a broker's license. The broker license exam was known to be difficult and required a 70% score to pass. One had to sign up in advance to take the exam because of the large number of people who applied. I applied to take the exam some two months in the future. This would give me time to prepare and review the list of exam questions that I had purchased. Shortly after I applied I received a call from the Department of Real Estate in Sacramento informing me that I had been scheduled to take the exam the next week. I said thanks but no thanks because I wasn't ready. The guy told me that it shouldn't matter because no one passes on the first try anyway. That made no sense to me. I told him that wasn't the way I did it. I was going to take the exam when ready and pass. He mentioned that I would lose the fee I had already paid. I rescheduled anyway and took the exam on the date I originally planned. They may have waived the second payment of the fee – not sure. This kind of exam is multiple-choice with four possible answers and has three different types of questions. One type of question tests for basic knowledge and most of those taking the test will get these correct. The next category of questions is more difficult and requires a higher level of knowledge. The third type consists of questions where the choice can be narrowed down to two of the answers. Both of these answers have a main part but also have added qualifiers. The answer that sounds the best usually has a qualifier that is weaker than the other or may even be incorrect so a higher degree of knowledge and even logic are required. A period of six hours is allocated for the test with a lunch break in the middle. When I took the test, I flagged all the questions where I was uncertain of my answer planning to go back as time permitted and give them some more thought. As I got to the end, I found that I had flagged about one third of the questions. Clearly, I needed to get a lot of these right. In the end, I recognized the pattern in these questions and changed most of the answers. The state agency doesn't tell you what score you got. It is pass or fail. I passed. I was now a licensed broker. The clock was now running. If I kept at it for a total of two years, I could open my own office and try to hire as many agents as I could to make money for me.

Napa was a great place to live. The weather was good even in the winter with most days mild and sunny. The drive up-valley to Yountville, Rutherford, and St. Helena on Route 29 was easy with light traffic. One of our favorite stops was Ernie's Wine Warehouse in St. Helena. Ernie's featured lots of low priced wines of good quality. My take on this was that there were a lot of start-up wineries in the valley that had not yet established a network for distributing their output and were using Ernie's to move their inventory. We were never at a loss for how to entertain out of town visitors. A trip up the valley for wine tasting worked really well. Carole and I reserved one morning during the week to play golf. Our favorite course was a nine-hole course that had a couple of par four holes and the rest par 3's. It was a relaxing environment with a lot of retired people as the regulars.

I thought a lot about sticking it out for another year and a half selling real estate and then opening my own office. Clearly that was the only approach that made any sense. I also thought about returning to the engineering field in aerospace or even taking a regular job in a new field. I started look-ing at the jobs listings in the newspaper. I answered a Digital Equipment Company (DEC) ad for sales representatives. They were big in business and scientific computing at the time. The interviewer conceded that I did know something about sales but no job offer. If I could find a good job in the local area or within commuting distance, we could stay in Napa. That would be good, but that seemed to be a long shot. I expanded my search and found that the job market was tight. After a few interviews, I concluded that I would have to focus on aerospace engineering where I had a lot of experience. I got the impression that companies were not all that eager to hire engineers over the age of 40, further complicating my search. It began to look like I was going to have a tough time.

Answering ads and working with the personnel departments is a tough way to go. They don't really know what the engineering organizations need and look for certain key words in resumes. I decided to get creative and wrote the scientist at the Boeing Company who had headed up the "Indian Summer Study" that I had participated in six years ago when I was working for GRC. I soon got a response, asking me to come in for an interview in Seattle. I spent a day with Fred, the manager I would be working for, talked with some project leaders and toured The Kent Aerospace Center and Boeing Field. They were definitely interested. It turns out that I had an excellent reference. The wife of the manager that escorted me around had a brother-in-law who I had worked with at Autonetics. He was in another group but we worked together to add the effects of real test data to the Minuteman accuracy estimates and we got along famously. It was now Halloween weekend. I was able to get a couple of other interviews and job offers from Vandenberg AFB and Lockheed in Sunnyvale. The Vandenberg group was interested in my Kalman Filter skills. This made me a bit apprehensive. I knew a lot about Kalman Filters but had never actually developed one on my own. With two job offers in hand, I called Boeing and told them I had other offers and needed to make a decision. They came through with an offer that beat the other two and said they wanted me to start the first week in December. Clearly, they needed me quite badly because companies don't like to hire in advance of the Christmas holidays and pay for all that vacation for someone who has just come on board. The Seattle area is known for its rain and that might seem like a drawback, but I had lived with my family in Snohomish, a small town just north of Seattle, when I was in elementary school and had positive memories of the area. Somehow the rainy weather had not made much of an impression on me. I accepted the Boeing offer.

Boeing gave us a house hunting trip before I even started work. Carole and I spent several days looking at houses in the suburbs of Seattle. One objective was to find a place that was not too far from any of the many Boeing facilities in the area. Our kids wanted us to get acreage so they could raise animals. There were a number of such places available but they were small, older homes that needed work. We decided to buy a house in the foothills a few miles south of the community of Issaquah. It was still under construction and on a short acre of land with tall fir, hemlock and cedar trees. On the last day of our trip, a severe winter storm came in. We were still dashing around in the snow storm making second visits to the interesting properties. Our plane was leaving at 5PM. We had to make our decision, sign the papers and get to the airport before the storm shut down the roads. Fortunately, our real estate agent had a vehicle with good traction and got us through the snows and commuter traffic and to the airport in time for our flight home.

I began commuting to Boeing. I got on an airplane on Sunday evening, worked 5 days at Boeing and flew back to Napa on Friday night. In the beginning, Carole drove me to the San Francisco airport to catch the plane and then met me on Friday when I returned. We soon discovered the Napa airport shuttle which had its starting point just a few blocks from our home, and I used that from then on. Boeing shuts down for the Christmas – New Year's holidays, and I had over a week at home. On New Year's Day, I started driving one of our 2 station wagons to Seattle so I wouldn't have to rent a car. The trip was uneventful until I got to the Siskiyou Pass between California and Oregon. There was a lot of snow and ice on the road and cars were stopping to chain up. I had my chains from past ski trips and quickly got them on. Some drivers had decided to risk it and had proceeded without chains. As I drove down from the summit, I saw a lot of cars that had crashed into the barriers and guard rails. I drove slowly and kept it on the road. Things were just getting interesting. In Oregon I ran into an ice storm that lasted all the way to Portland. I had to stop every few miles and clear the windshield. In Portland, the challenge was getting across the Fremont Bridge on the Willamette River. The bridge had iced up and cars were sliding around. I had removed my chains earlier and didn't want to put them back on so I drove carefully and without an incident. Once In Washington, the weather was warmer and the roads were good though wet.

By the end of January, the new house was supposed to be finished and it was time to move the family to Washington. We decided to keep the house in Napa and rent it out. The real estate firm I had been with managed rental properties (hoping to be able to sell them at some point), and I turned it over to them to handle. The movers came once again and collected our possessions. We packed up our other station wagon with what we would need for a few days and also our cat, dog and a hamster. We said goodbye to Napa and headed north. Unfortunately, the hamster did not survive the rigors of the trip. The dog and cat were fine.

The Boeing Company - Seattle

Boeing Aerospace

I JOINED THE Guidance and Navigation Group in the Boeing Aerospace Division, a staff organization of engineers with skills in the guidance and navigation disciplines. The manager of this group, Fred, is the one who escorted me around on my interview trip. My first assignment as a Boeing Co. engineer was to support a proposal effort that had recently gotten underway. The proposal team had office space at the Developmental Center that was located at the southern end of Boeing Field in Seattle. This was an exploratory effort to decide whether to go ahead and participate in an actual proposal to the US Air Force. The purpose of the project was to develop a system that could destroy Soviet satellites. Surplus Spartan missiles from the old Sentinel System would be used as the interceptor missiles. My role was to define the intercept guidance and maneuver strategy and develop requirements for performance.

For the engagement scenario, a ground-based radar would track the satellite that was to be intercepted. A Spartan missile would be launched and radio guided onto a path that would come close to intercepting the target. As the Spartan neared the satellite, its terminal sensor would acquire the satellite target and began measuring azimuth and elevation angles to the target and compute angular rates. The terminal stage sensor could be either radar or infra-red (IR). The on-board guidance algorithm would issue commands to the engines to null the azimuth and elevation angular rates to zero, putting the interceptor on a collision course with the target. The satellite would be met head-on, resulting in a closing velocity of over 3 miles per second. This head-on geometry is necessary to eliminate the effects of any timing errors which would pose an accuracy problem if a crossing trajectory were used. The only other operation left to execute would be for a terminal range sensor to determine the right time to detonate the warhead. The warhead was to be a drum containing hundreds of tungsten rods which would expand in a shotgun pattern. At 3 miles per second, any impacts with the target would result in gaping holes sufficient to disable the satellite.

I was assigned a programmer to write the code for the guidance algorithms and a simulation of the engagements. I was once again an engineer and "more than just a programmer." My programmer had a PhD in mathematics/topology. That seemed to me to be a particularly useless field of knowledge. He had been teaching in the university world and had become disenchanted. His uncle was a mid-level Boeing executive and he may have used that connection to get hired. I got the impression that he looked upon his current assignment as a stepping stone to something better; in fact, I began to sense that he wanted my job although he clearly did not appreciate the scope of my knowledge and experience. I made an effort to keep him in the dark although under a different and a more constructive attitude on his part, I would have been glad to teach him what I knew.

I concentrated on modeling the engagement and computing the angular rates using my curve fitting tool from Safeguard days. There was still a lot of work to do including a proper error analysis of interceptor terminal stage performance to determine instrument accuracy requirements although much of this would probably be done after the contract was awarded. The use of a Kalman Filter to track the satellite and the interceptor and to compute miss distances had been mentioned. I said that I would do that after we had won the contract. It was now time for me to actually construct and check out a Kalman Filter just to make sure I could do it. I started staying late after working hours to write a computer program with an actual filter. This proved to be a relatively easy task and I had it working after only a couple of evenings. There are a lot of good books available on the subject. The one I found to be the best is "Applied Optimal Estimation" edited by Arthur Gelb and written by the technical staff at The Analytic Sciences Corporation.

I moved my wife and children to the area at the end of January. It was time to do so, as the children needed to start school at the beginning of the semester. Unfortunately, the new house was not ready and we spent three weeks in the Holiday Inn in Bellevue. The motel was understanding enough to let us have our dog and cat in our room (the hamster didn't survive the trip). We finally moved into the house and the children were enrolled in the Maples Hills Elementary School in the Renton Highlands, David in fifth grade and Cheryl in third. We were out in the country, about seven miles from the town of

Issaquah, which was our mailing address. The neighborhood was known as Mirrormont (from Mirror Mountain, I believe) and contained many houses similar to ours, four bedrooms and on an acre of tall trees.

Boeing decided to go ahead and write a proposal for the project I was working on and brought in a new proposal manager, Jerry, to show that they were serious. Jerry was a veteran of many such efforts and immediately showed he was in charge. The previous manager was doing a reasonable job but lacked Jerry's dynamic personality. I continued to work on my part and refine my results. My supervisor was a long term Boeing engineer and had experience on many projects but didn't know that much about what I did. When I had worked at Autonetics, I was used to having a boss that understood everything I was doing. This was not going to be the case at Boeing. He was also a newly licensed private airplane pilot and pressured me to go flying with him during the lunch hour. I wasn't interested as I knew that accidents in private aviation were much more likely than with commercial flying. But he persisted, and I reluctantly agreed to go with him. We circled around in a single engine Cessna for a while and landed without incident. I hoped that would do the trick, but I was wrong. He wanted to go flying again a few days later. I had to give him a firm no. It seemed to hurt his feelings when I told him I thought it was too dangerous.

The proposal was turned in, and now it was time to wait. It would take a few months to find out whether we had won. There was a project at the Kent facility called the IUS (Inertial Upper Stage) that was behind schedule and needed bodies. I was drafted to work on it. The IUS was a two-stage solid-fueled rocket upper stage being developed by Boeing for the US Air Force to raise payloads from low Earth orbit to higher orbits following launch aboard a Titan rocket or from the payload bay of the Space Shuttle. The group I joined was developing computer code to manage various aspects of the mission. The program was behind schedule and over budget. A new program manager had just been appointed to try to turn things around. Shortly after coming on board, he announced that there would be a ten percent cut in the number of people working on the program. This seemed strange since we were supposedly short-handed. In later years, I learned that this strategy could be effective because it incentivizes people to buckle down and get to work and weeds out the non-contributors.

Working conditions were far from ideal. About 100 to 150 of us engineers were placed in a large room with desks arranged end-to-end in long rows and only 3 or 4 feet between rows. The noise level made it hard to concentrate. I noticed from the beginning that there were a number of engineers that weren't doing anything productive. One such individual, an older fellow, sat in front of me and often turned around and made various remarks that had a double meaning which gave me the impression that he was some kind of pervert. He had a PhD in mathematics and should have been able to contribute something but didn't seem interested in doing so. There was an attractive young lady seated at the front of the room. She had some accounting or administrative function. She usually wore tight clothing or short skirts, and every hour of so she would get up and slowly walk to the back of the room where the coffee pot was located. All eyes would follow her on her mission to fill her cup and then escort her on the return trip. It would be hard to estimate the number of hours of work that were lost each day.

I was designated to be a "lead engineer" and assigned another engineer to supervise. He had been drafted from the B1 Bomber program along with a buddy. They both complained frequently about how they didn't want to be here. I assigned him a trial task to see what I was dealing with. The job was to write a FORTRAN subroutine to print out several different kinds of data the computer program was calculating. The main program was clean and consisted of 4 or 5 call statements, one for each of the functions it performed. Instead of adding an output subroutine, this clown cluttered the main program with numerous write and format statements so that someone looking at it would have no idea what it did. I couldn't decide if he was that inept or he screwed it up on purpose. I informed my supervisor about what he had done and advised that he be sent back to B1 since he was of little or no value.

Companies are known to underbid a contract in order to win and then recover by aggressively charging for any changes that the customer, the Air Force in this case, wants to make. We were asked to provide some added ground support functions and write a number of documents to support them. My supervisor asked me to scope some of this work and estimate the man-hours needed including the number of pages of documentation and the effort required to produce them. Clearly, I was no expert at this task, having never done anything remotely close to it, but I did my best and came up with some numbers. My supervisor looked at what I had done and doubled my estimates. These were then sent to the contracts department where, I heard, they were doubled again. Boeing was going to make some money with the IUS Program.

The guidance approach was a challenge because of the two solid boosters. Separate rocket burns had to be managed so that the payload was placed in the desired orbit. One of the guidance engineers came up with an ingenious approach to achieve this. Additional flexibility could be added by adjusting the fuel formulas. There was an executive guidance manager assigned full time to IUS. He was something of a "godfather" figure in that he interviewed and approved any guidance related staffing on the program. He was also tied to the guidance and navigation staff organization of which I was a member. He was often consulted on the placement of guidance personnel anywhere in the division, so it was important to be on the good side of him. I, of course, didn't fully appreciate his role and importance until sometime later.

After a few months, I was becoming a reliable member of the organization and one who was actually cranking out some useful stuff. However, the word came out that Boeing had won on the proposal I had worked on when I first joined the company. I was contacted and asked to join the team just being formed as the guidance and accuracy expert. This information was passed on to the godfather and I was called into his office. He told me that I could choose either of the two assignments and that the decision was up to me. I saw this as a lose/lose choice for me. I thought that the powers that be should have made the decision that was best for the company. Why were they passing the buck? IUS was something of an disorganized mess with a lot of marginally qualified people on it. I saw the situation as an opportunity to escape IUS, but I suspected that the godfather would be angry if I did so and my choice might hurt me later on. If I stayed, I would be passing up a great opportunity to do some real engineering. In the end, I reasoned that my sanity was more important than IUS so I bailed. I wished them well.

The project was given a cryptic name, the Conventional Launch Segment. What we were trying to do was apparently politically sensitive. Also, some aspects of the project were classified. The system was going to be based in the Kwajalein Islands and use the s-band radar there for tracking the targets and also the interceptors. This radar is accurate in range but the angle data is much less so. Two other s-band radars would be built on neighboring islands and with three good measurements of range, a very accurate track could be established. The three radars had to be netted and a very accurate time reference would be required. The technology to do this was said to be available.

My role was to determine the accuracy that could be achieved at intercept. This would be used by the lethality analysts to determine kill probabilities. I spent some time with the location geometry of the three radars to see if I could come up with a closed form solution of position, but concluded that it was a difficult task. Fortunately, the Kalman filter I was developing did not need this. We were looking at both radar and IR as terminal sensors. Another engineer was brought in to work on one of these. He was well qualified and we worked well together on these options. I was asked about using the same programmer as I had used during the proposal and said, "No." His uncle called the program to find out why he hadn't been picked up. My rationale was that he was overqualified. That seemed to satisfy everyone. The job certainly didn't need a PhD. The programmer that was selected to assist me was qualified and seemed content and happy just to do what I asked of him. Jerry, who had led the proposal effort, was named project manager and he immediately set to work inspiring and motivating us. He liked to have monthly reviews and frequent show-and-tell meetings with the Air Force. This really caught our attention and forced us to focus on our tasks. Jerry's management style made the project the foremost thing in our minds.

The target tracking Kalman filter development went smoothly. I had the Spartan trajectory model from my days with GRC in New Jersey and this was added to the simulation. The intercept scenario was to first acquire and then track the target satellite. Once the satellite was in track and an intercept point calculated, the interceptor would be launched. The interceptor was command guided from the ground and placed on a near collision trajectory to hit the target. Tracking was switched from the target to the Spartan as it neared intercept in order to get an accurate track on it and make adjustments to its trajectory. The sensor on the terminal vehicle would then acquire the target and make the final adjustments. A time-to-impact sensor would determine when to deploy the warhead. This was all soon up and running and I was generating miss distance statistics by running the simulation in Monte Carlo mode (multiple statistically independent runs). These data were then handed off to the lethality group to compute kill probabilities.

One of our tasks was to come up with a strategy for dealing with large objects (aka extended targets). There is a phenomenon known as glint or scintillation associated with them. There was a radar manager on the project who had one person assisting him. He was probably hired because of his experience working on the TRADEX radar on Kwajalein. Unfortunately, he didn't seem to have any analytical background and tried to work the glint problem by searching for information in the existing literature and trying to identify anyone who knew something about the subject and would help him. This strategy wasn't producing anything, and I needed input to evaluate the accuracy that could be attained tracking the extended targets in the threat. The glint phenomenon occurs for large targets because such targets contain a number of scattering centers and returns from these interact, either amplifying or attenuating the return signal, and result in range noise. I decided to simulate the effect using a 2 center model and rotating them. This produced some interesting results which we presented to some people at Lincoln Laboratories MIT who were consulting for the Air Force on the project. They seemed impressed and encouraged us to continue work on the model. However, we didn't have much time to come up with something and creating a more realistic model for the extended targets seemed like a major undertaking. We had received some actual extended target tracking data from MIT that showed range measurement noise over several minutes of track. I sampled this data at one second intervals and used it in my simulation to model the range noise. It was in a continuous loop and was initialized with a different random number for each of the three radars so that noise statistics were uncorrelated. This was soon recognized as a good way to model the glint phenomenon and was accepted by the project manager. The radar manager was upset that I had solved his radar problem for him, but recovered enough to frequently claim credit for it by claiming that "we" had solved the problem. He did deserve some credit, I suppose, for finding the MIT data.

With the extended targets problems behind us, I was able to provide the lethality group with miss distance statistics for all of our targets in the form of xyz position error matrices. They wanted these in a coordinate system that had one axis along the relative velocity vector, so I transformed my results to this system. They announced their lethality results at one of the periodic briefings with the Air Force. Kill probabilities for the lower altitude engagements were in the 90% range, but there were some missions where kill probabilities around 70% were being predicted. Something was clearly wrong as there was no obvious reason for this. The group had several people working on the problem, and they seemed to be mining the minutia to improve performance. One Air Force observer was quoted as saying, "They're polishing a turd." I decided to look into this even though it was not my responsibility. I didn't want the program to fail. I found that they were timing the warhead release so that the radius of the pattern of deployed rods would be approximately twice the expected one sigma miss distance, thereby ensuring impacts with the target in a high percentage of the cases. This worked well for the targets with low anticipated miss distances, but for the extended and high altitude targets, the pattern was so thinly populated that the 2 or 3 impacts needed to produce a kill didn't happen, resulting in a high proportion of misses. The answer was quite simple: keep the pattern of rods small enough so that any target falling within it would be destroyed. The probability that targets that would fall outside the pattern was statistically small in comparison, even in the worst cases. I discussed this finding with my boss who reacted by saying, "How are we going to tell them?" I was finding out that in the Boeing culture, one doesn't help with or interfere in what someone else is doing no matter how screwed up it might be. I decided to write a memo describing the situation and distribute it to the boss, the chief engineer, as well as, the lethality group. At the next review, the kill probability numbers were much improved.

The Air Force team included some consultants from the Aerospace Corp. One of these was a PhD who was originally from Argentina and had a rather thick accent. We didn't hear from him very often, but one day it was announced that he was going to make a presentation that all of the engineers should attend. He addressed a problem with handling data he described as an ee-lip-suh (i.e., ellipse). He seemed to be talking about the miss distance statistics that I was generating. He drew an ellipse on the board and stated that if the major and minor axes were rotated about an axis coming out of the board at the center of the ellipse as in a coordinate transformation, some data would be lost because the intersections of the rotated axes and the ellipse were clearly smaller in magnitude than before the rotation. He proposed a way to make a correction that would adjust for this effect. It seemed to me that all he really needed was a simple coordinate transformation. Also, he was only dealing with a 2 dimensional ellipse. The miss distance data was 3 dimensional and therefore an ellipsoid, but he never addressed that. My supervisor was in the meeting and seemed impressed by Dr. Ee-lip-suh's presentation. The doctor had never talked to me about what I was doing or how I was doing it so I had to wonder where this was coming from. After the presentation, my supervisor called me into his office and asked me if I was going to follow Dr. E's method. I said no and that I didn't think Dr. E's method had any merit and, furthermore, the way I did it was correct. He seemed stunned and just looked at me. He called in the other guidance engineer on the program and asked him if my approach was right. He said that what I was doing was the right way. I thought Dr. E would follow up, and I was prepared to give him a lesson in coordinate transformations, but he never showed up and the subject was never brought up again. Someone must have gotten to him.

To generate miss distance statistics, I ran my simulation in Monte Carlo mode with 20 or so uncorrelated runs of the same engagement. Each run would produce an xyz calculation of miss distance as a 3x1 matrix. Multiplying this matrix by its 1x3 transpose produced a 3x3 matrix of squared errors. By summing these matrices and dividing by n, the number of cases, I produced a matrix of one sigma errors and cross axis correlations. These were computed in tracking coordinates. To transform this data to the relative velocity coordinate system favored by the lethality group, I pre-multiplied the error matrix by the appropriate transformation matrix and post multiplied the result by the transpose of the same matrix. This was all pretty basic. Maybe I should have given a seminar on how to do this to build trust among those who were encountering this method for the first time.

I got a phone call one day from one of the Air Force lieutenants on the program asking if he could come by and talk to me about my tasks. I agreed and we had a conversation about a few issues. Word of this meeting got back to Jerry, the program manager. He went ballistic. Informal contact with the customer was definitely prohibited. Everything was supposed to go through the project manager. He was right, of course. Lesson learned: A good manager wants to and needs to know about everything that is happening on his project and be the sole source of any information coming from the project.

As the project began to wind down, it was clear that we had achieved our objectives in defining a system that would meet the performance requirements. It was up to the customer to decide if we should go ahead and build the system. Our competition was a Vought Corp. system based on launching an anti-satellite missile from an F-15 aircraft. Our system was vulnerable to attack. It would be fairly easy to take out some of the critical components, such as the radars. In the end, perhaps for this reason, the Air Force went with the Vought system.

After the CLS project ended I returned to my home organization. The Global Positioning System (GPS) was a new technology that Boeing wanted to learn more about and keep up with developments. GPS conferences were being held every few months, and I was sent to these to educate me in what GPS was all about and to show that Boeing was interested in being involved in the program. I acquired all of the available literature and read it. Several companies had programs to develop GPS receivers. One of these was Texas Instruments. TI was looking into using real time GPS position and velocity data to update the position and velocity of an inertial navigation system. Boeing contributed some $50k toward a TI test so we could be involved in the experiment and share in the data. I went on two trips to TI to participate in the activity. TI had built an airborne GPS set that was contained in two large electronic boxes and weighed about 70 lbs. total. We loaded this onto an airplane and took off. The test didn't produce any useful data for us, but it was a start.

There was a small advanced projects group in the next building that had some DARPA funding and was in the business of looking at new ideas and future systems. The manager of this organization (Jim) had been a defensive tackle on the Washington Huskies football team during the Hugh McElhenny era. Everyone knew about his football career, and he seemed to have gained some status from this. He occasionally made mention of his football days and implied that he and McElhenny had some differences. This seemed to further enhance his image. He claimed to have several high level clearances that "I can't even tell you about." His group had a contract with the Rand Corp. to study GPS based guidance and navigation of reentry vehicles and a satellite system. An engineer from my staff organization had been assigned to do the guidance work but did not have the tools or knowledge to do so. I was assigned to replace him. I set to work to create a simulation of the GPS system and put together a usable tool in a couple of weeks. I had a ballistic missile simulation in my tool kit and connected GPS measurements to the position and velocity of the reentry vehicle using a Kalman filter and soon began producing results. Jim told me that if I came through on the guidance end, he would see to it that I got a $10k per year raise. I doubted that this was possible, but it was a nice thought. A Rand analyst had predicted some wildly optimistic guidance accuracies based on a covariance analysis he had done. This kind of analysis is known for sometimes being too optimistic and his results were clearly so. The Boeing team (including me) made a couple of trips to the Rand Corp. in Santa Monica to present our results. They liked our conclusions and funded our continued research. There were some classified aspects to the project. It started out unclassified then went to secret and then top secret. I applied for a TS clearance, but this takes time. For a while, I was not allowed to work on my own program.

Jim asked me to identify all of the available ICBM guidance systems, including those in development and compare accuracies. I looked at the Delco USGS (Universal Space Guidance System), the Minuteman NS20, the Singer-Kearfott stellar-inertial system and the AIRS that had been selected for the MX missile. I wrote a memo with my results. The stellar-inertial system seemed to be the best based on cost and performance; however, in my experience, S-K was very aggressive and optimistic in predicting future performance. There were other groups at Boeing who were looking at future missile developments using inertial guidance and I discussed my findings with some people from one of these groups. A few days later, I got a phone call from the head of Boeing's Minuteman program. We had a difficult conversation. I think he just wanted to know "Who the hell is this guy?"

I was on occasion told to drop everything and help someone with a problem. A project manager, who was leading a project to develop a rocket artillery system for the US Army, had called Fred and asked for assistance in applying statistics. The system had been test fired and performed well except for a couple of bad misses. He had his staff do a statistical analysis that might explain what had happened. They had come up with nothing (at least not what he wanted), and he concluded that they didn't know what they were doing. I went over to the project and talked with the manager. He had been trying to work the statistics problem himself and seemed to be on the right track. This was not my field but I had taken courses in statistics, had some experience in the field and could read a text book. I spent a day or so working on the problem and made my recommendation that what he was doing was the correct approach and he should continue. He was trying to come up with a rationale using statistics that would allow him to exclude the bad misses from his accuracy statistics by claiming they were "outliers." Statistics can be a tricky business. The probability that the misses could occur as part of the expected distribution was very low. One might be able to exclude one bad miss, but both? That seemed rather "iffy." Perhaps the missiles had malfunctioned, which is an explanation that might well have been the case and might have helped him. A few days later, the manager called my boss and extravagantly praised my efforts. I think he wasn't quite sure that what he was doing was OK and just wanted someone to tell him he was on the right track.

When I was between assignments once, I was put on a project to research a "rail gun." This was an electromagnetic device that would be used to propel an object into orbit. One stated use was to find a cheap way to get rid of radioactive wastes by launching them into space on a trajectory into the sun. I researched guidance system components that could survive in a high-g environment. The accelerations for a rail gun would have to be huge. To achieve a velocity of even 10000 fps over 1 second would require a rail 5000 feet long. To reduce the rail to only 50 feet, the device would have only .01 seconds to achieve the desired velocity. I didn't know much of the physics, but it seemed unlikely this could be done.

My home organization boss, Fred, invited me to attend a meeting with some managers from a Northrup Company division in Rhode Island and from Rockwell-Collins in Cedar Rapids, IA. I was asked to prepare a slide showing an integrated system containing a GPS receiver and an inertial guidance system. I presented a very simple diagram. It was well received although I got the feeling that they believed that there was a lot of thought behind my five minute block diagram. This doodle was to become the basis for a joint project between Boeing and these two companies. A follow-up meeting was scheduled to occur in two weeks. I received no direction from my boss on what my role in this was or what, if anything, I should be doing to support this activity. I thought about it for a while and realized that I was the only engineer/worker involved. It looked like I needed to take responsibility for moving the agenda forward, so I expanded upon my earlier diagram and added some data transfers between modules. This went over well and became the basis for the three company joint project. Fred informed me that I was going to be the project manager for the Boeing side of the project although it was clear that he was reserving the right to jump in and take over at any time.

This arrangement resulted in some embarrassing moments for me. Boeing zealously guards the management designation and is reluctant to create any new managers and provide that which goes along with it. For this project, I should have been provided with an office and treated as if I were a manager. But alas, all I had was a desk in the bullpen. When the project meetings were held at Boeing, I had to schedule a conference room from which to conduct business. I got no help from my management. It was up to me to figure out how to run my end of the project. One of the representatives from Northrup, a marketing manager I think, noticed that I was inexperienced at the job and told Fred that he wanted someone else to represent Boeing. Fred gave me the news and said I would continue on with the project, doing the engineering and other heavy lifting, but I wouldn't be the project manager. A figure-head manager would be appointed to attend meetings. I said no thanks to this arrangement and stated that I wanted another assignment knowing full well that he didn't have anyone else who could do what I did. Fred appeared to think this over for a while and then told me I would stay on as the manager and that he would inform the Northrup guy of this. He did appoint a Boeing shadow manager for the program. This guy began showing up at meetings but didn't participate. I can only guess what assurances Fred had given the Northrup guy to make him think all was well.

Northrup was sending a vice president to our meetings and Collins sent an upper level manager. I, a lowly engineer, was representing Boeing. Fred wanted to get some recognition for this program with an article and photos in the Boeing News, but having me as the Boeing person leading the program was not going to cut it. A Boeing vice president was recruited to appear to be involved in the program. A photo was taken of the Boeing and Northrup VP's along with the Collins manager and published in the Boeing News along with a brief description of the "joint project." The Boeing VP had zero involvement in the program. I was good enough to do the engineering work and define the program concept but not good enough to get the credit.

Eventually the other companies sent some people who would be involved in the work. In one of our first conversations, someone from Collins mentioned interface control. Bingo! It hit me that this was the key to the success of the project. An interface control document is something I had heard Boeing systems engineers talk about and realized this was a critical document that I needed to prepare. I put together a draft identifying the data that would be sent and received by each module, these being the Collins GPS receiver, the Northrup inertial guidance system and the Boeing integrating computer. All data items were identified as to type, format and frequency of transmission. I put together a simulation of the entire system. I already had a model of the GPS system and added a simulated IMU and an airplane in flight. I also created a Kalman filter to use the simulated GPS measurements to update the IMU position and velocity. I had 2 to 3 engineers working for me on the project. I assigned one of them to get the Kalman filter working. He was a junior person and struggled to accomplish his task. After a long period of not getting anywhere, he concluded that my Kalman filter was defective and would never work. I had to spend a day finding the problem and getting it working. My Kalman filter was fine.

The project required a lot of travel. I took several trips to Rockwell-Collins in Cedar Rapids, IA. These were initially get-acquainted meetings for the teams and then progress updates as things evolved. The project had a lot of visibility at Boeing which meant that a vice president accompanied us on some of the trips. There were also a couple of trips to the Northrup facility in Rhode Island.

Collins was getting ready to deliver the GPS receiver which meant that the integration task at Boeing needed to get underway. The Guidance and Navigation group maintained a hardware laboratory where the integration activities would take place. I found a bright young engineer who worked in the lab and seemed to be able to handle the task. I recruited him for the program. Our first conversation was interesting. He seemed to expect that I would give him specific instructions on exactly what to do. I explained that I looked to him as the expert and he needed to figure out what to do and how to do it. He seemed to like having that responsibility and immediately jumped in and began making progress. My plan was to move my simulation code over to the VAX computer in the lab and replace parts of the code that simulated the hardware with the real hardware as it became available. My engineer transferred the code and quickly got it running. The interface control document that we had written provided the information specifying what data would be provided and the format of the data, making the integration task straight forward and relatively easy. Access to the lab was controlled so I wasn't free to come and go, making it difficult to get involved in the integration activity. Before I could work this problem, Fred pulled me off the project to work on the new small ICBM (SICBM) that was just entering development.

My supervisor recognized the fact that I would continue to be in a management role of some sort now and in the future and enrolled me in the Boeing after-hours Pre-management course. This course wasn't so much about how to manage people but emphasized the administrative functions a manager performs. I realized that I needed some real management training and took the initiative. I enrolled in an MBA program that was given by Southern Illinois University at Edwardsville and sponsored by the McChord Air Force Base in Tacoma where classes were held. It was a weekend program where one attended 20 hours of classes every third weekend. Classes were 6 weeks long and one could take 2 courses per quarter by going "half time." I had taken a lot of engineering classes at my previous employment locations, but had never completed the requirements for an MS or MA degree. I didn't see much value in an advanced engineering degree at this stage of my career as highly specialized as it had become. But an MBA where I might learn about management seemed like a good idea. Fred was lukewarm to this endeavor at best and probably was somewhat hostile to it. Boeing managers tend to have no formal management training and learn from "on the job training." This results in a fairly limited scope of management expertise. Boeing did have an excellent education support program and generously paid for all my tuition costs. The program was supposed to take 3 years to complete with half time attendance. When one enrolled in a course, one was provided with a text book, various articles to read and a homework assignment. After 3 weeks, the class met at McChord AFB with a professor/instructor who had been sent out from Edwardsville. We met for 4 hours on Friday evening, 8 hours on Saturday and 8 hours on Sunday. There would be a test at the end of the Sunday session. The key to success with this format was to begin studying right away and to allocate enough hours for study each day so that when it was time to meet, the book had been read, the homework finished and any required reports written. I found that 2 or 3 hours of study each evening were adequate to accomplish this.

The first class was almost a disaster for me. The students who, like me, did not have an undergraduate degree in business had to take as many as eight undergraduate courses. One of these was Management 101 and the first class I attended. The instructor was a character, to say the least. He ignored the text book, said we shouldn't have bothered with the homework and launched into a story about the business, a bar and restaurant, he had established in St. Louis. He spent the entire lecture periods talking about how he set up the business including selecting the location, hiring the staff, picking the logo and other such details. At mid-term there was a short essay test "because the school makes me do it." It became clear that grading in the class was going to be highly subjective. When one of the students made a comment that he really liked, it seemed clear from his reaction that the student was going to get an A. In the second lecture session, he continued with his dissertation on his restaurant and then gave us a short final exam, again in essay format. I ended up with a B grade and no real reason why. I did read the book so I learned something. Some of the professors who came later mentioned that our instructor was under disciplinary action for his flakey behavior and also that his restaurant was failing due in large part because the employees were stealing and "robbing him blind." I suppose this could be called justice.

Fortunately for us, the instructors who came later were all competent and some were even very good. The program was not supposed to require much mathematics but, of course, a certain amount was required. Some of the students struggled in this regard and a few even dropped out of the program because of difficulty with the little math that was needed. For me, the math was easy and fun and gave me an advantage over most of my classmates.

There were several team assignments, but we were most often paired with one other student. This, I suppose, was a good format that taught us how to work with others. One of the courses was a small business development exercise. In the first half of the class, each student was charged with putting together a small business plan for the business of his/her choice. I chose Computerland, which was a hot franchise at the time for setting up a retail computer store. At the midterm meeting, the instructor selected a few of our plans that showed the most promise and put us in teams of 5 or so students to further develop the plan and present it at our next meeting. He selected my plan, Computerland, as one of these. We held a team meeting and assigned a piece of the work to each member. We met a few days before everything was due and each team member handed in his work to me and my usual team partner, Raj, to put together and create the classroom presentation, except for one team member. He passed us a few meaningless numbers on the back of a sales receipt and said he would have his real results in a few days. We gave him a little more time, but it soon seemed clear he wasn't going to produce anything. Raj and I met and each took half of the recalcitrant team member's task and put together the team presentation materials at the last minute. We learned about working with a larger team and also took away a valuable lesson, that being that sometimes team members don't pull their weight. One needs to be aware of this and be ready to step up when it happens. Anyway, our presentation was a major success and I received an A in the course. Raj let the instructor know about our teammate's failure to perform. I'm sure he got marked down for that.

We had to write a lengthy report for many of the classes. Some were team assignments, but most were an individual effort. I wrote my reports in longhand. My wife, who is an excellent typist, typed them for me. This saved me a lot of time. I was surprised to find that a number of my classmates were not turning in their reports at the end of the class. They were allowed to turn them in late but would receive a reduced grade.

Near the end of the program, we were required to take a class called Corporate Policy. It was billed as a summary class, drawing on much of what we had learned. The instructor was a senior professor who other professors had described as "arrogant." He hadn't been mentioned by name, but we knew who they meant. He and I got off to a bad start. I tried to impress him by asking intelligent questions that indicated that I knew quite a bit about the corporate environment. He clearly didn't like that and I received low grades on my written work. Being aware of the arrogance tag and after considerable thought, I changed my strategy and decided to ask naïve questions that gave him an opportunity to expound on the topic and to show his great knowledge and insights. This worked like a charm and things were looking up as we approached the final exam. The final would require each student to read a company case history, analyze it and write a report describing why and how the company's actions created their current situation. There was short list of companies, but we didn't know which one it would be. When I got my test, I found that the company was International Silver, a company our investment club had owned when I worked at Autonetics. I knew this company quite well. IS had a lot of cash in the early 60's and embarked on a program of acquisitions. This increased their earnings growth and the stock price soared. After using up their cash resources, they continued to acquire other companies but with borrowed money. Earnings slowed due to the cost of debt service even though sales increased. Consequently, stock performance suffered. My insights into IS were probably due to my investment experience as much as anything I had learned in the MBA program. After the test, I was sure I had aced it. I got my A.

I got an A grade in all of my classes, except for that disastrous first class. Near the end of my program, I called the instructor of that first class and asked him if there was anything I could do to raise my B to an A. He said that yes, there was and gave me a long list of research to do and reports to write to accomplish this. Here was a guy who asked nothing from us when he was teaching the class and now he wants a huge effort from me to raise my grade. I never got back to him but gave him credit for his nasty and ironic sense of humor.

I finished the program and got my MBA degree in the expected 3 year period and was eager to apply what I had learned. There is a difference between education and experience and a combination of the two is what is required to be effective at managing. I was ready to get started applying what I had learned.

The three years spent working on my MBA degree took a lot of time away from the family but they seemed to do all right. There were some problems, however. David had just finished his sophomore year at Issaquah High School and had gone from being a top student to one who was struggling in all of his courses. He had grown his hair long and on the last day of school was pursued by a pack of "jocks" who vowed to cut off his hair. David informed me that he was not going back to that school. I wasn't exactly thrilled by his performance there either and realized that it was time to try something else. We found a private school called The Stanford School in nearby Redmond that claimed to be able to motivate students who were not performing up to their potential. Cheryl had just finished the eight-grade in Issaquah and had gotten good grades while there. She didn't want to go to high school in

Issaquah because there were a lot of out-of-control, aggressive boys in the school. She and her mother shopped around and checked out all the nearby options, finally settling on a Catholic girl's school called Forrest Ridge. Neither my wife nor I were Catholic, but it was a prep school and had a good reputation. Sending both children to private school was going to cost a bit of money. Carole had started a window coverings business working out of our home and was making some money, although it mostly went to fund her IRA.

The Air Force was developing a new missile, the Small ICBM (SICBM). Boeing was gearing up to bid on some of the contracts that would be awarded and was provided with some funds to participate in the systems definition studies. My supervisor pulled me off my GPS aided inertial project and assigned me to the SICBM effort. I moved to some office space at Boeing Field and began studying the latest developments in inertial guidance systems. I enjoyed walking to work in the mornings along the flight line and passing under the wings of an orange painted 747 parked just across Marginal Way and opposite the Boeing headquarters building where I worked. The aircraft was painted with the Braniff Airlines logo and was probably one that had been repossessed after the airline was temporarily shut down in 1982. My supervisor, Fred, worked with me for a few days at this location. One day he was talking rhetorically and eloquently about his success in getting some funds for unscheduled raises for the engineers in the group. Whoops! He then put his hand on my shoulder and said, "But you didn't get one." I thought, "What the hell was that about?" He certainly knew how to inspire me.

I had hired into the group with a salary that put me near the top of salaries in the group. Each year, a chart was produced that showed the annual salaries of everyone in the group versus their years of experience. No names were shown, but we could usually guess who was who. I think that the production of this chart was a requirement negotiated with the union that represented the Boeing engineers. The union is called the Seattle Professional Engineering Employees Association or SPEEA. I was always number 3 or 4 in salary. Boeing negotiated a new contract with the union every 3 years and usually offered what amounted to a cost of living raise. The union often talked about a strike but never did go out while I worked at the company. Over the years that I was an engineer, my salary increases almost exactly matched the amount of inflation. I had a bit of success in recruiting engineers to join the union and signed up as many as ten during my time as an engineer. I received a cash reward from the union for each new sign-up, which I used to take the new member out to lunch. My boss recognized one member of the group each year as the "Top Contributor." I won the award 2 years in a row for my work on the GPS-inertial navigation project. Perhaps this was in lieu of a real increase in my salary and/or a promotion to management.

The main contenders for the SICBM inertial guidance system were the Singer stellar-inertial system, the AIRS (originally developed by MIT Draper Labs and later transferred to Northrup) and a ring laser gyro system being pursued by several companies, notably Honeywell. The Singer system had a proven track record, but the company was making very optimistic claims for future performance. In their defense, the technology was advancing rapidly and claims that seemed unreachable were often later surpassed. The AIRS was taking a long time to develop and was very expensive. The ring laser gyro systems were new and did not yet seem to have the accuracy potential of the AIRS. I made several trips to Honeywell in Clearwater, Florida, Singer in New Jersey and Northrup in Torrance, Calif. to gather information on their guidance systems.

I had a number of conversations with a supervisor at Northrup about progress on their development of the AIRS. This supervisor had one of those gender ambiguous names (like Chris) and I couldn't tell from the voice whether I was talking to a man or a woman. I had even asked another Northrup employee if the supervisor was a man or a woman, but he simply ignored my question. I found this puzzling, but it was soon resolved. There was a contractor meeting at Honeywell in Clearwater, FL that I attended. As I entered the meeting room, I spotted an old friend from Autonetics and sat down next to him. He was eager to tell me about the Northrup supervisor who had formerly worked in the same group as he did. One day this guy had shown up in a dress and high heels, apparently in the early stages of his sex change. Later in the day, I introduced myself to the Northrup supervisor and we chatted briefly. The supervisor had long hair, was wearing a dress and appeared to me to be a mannish looking woman. I didn't know where she was in her transition, but she must have been fairly far along. Our interactions with Northrup were essentially at an end at this time, and this was the last contact I had with the Northrup supervisor.

Autonetics was in possession of some classified documents on accuracy projections for the AIRS and the stellar-inertial system. They were willing to share this information with Boeing but the documents would have to be viewed at their facility.

I boarded an early morning plane, flew to nearby Ontario and rented a car. As I entered the building where my contact, the MX chief engineer, had his office, I encountered a former colleague who had become the head of the MX Program. We had a short, pleasant conversation. He then escorted me in and to where my contact was. I spent all day poring over all the pertinent material and taking mental notes. Everything was classified "secret" and, therefore, I couldn't write anything down. I tried to commit to memory all the key numbers and then caught the evening plane back to Seattle.

We still had our rental in Napa. We had specified "no pets" but found out from our former neighbor that the tenant had 2 large Dobermans. I stopped by to look at the property on one of my business trips to the Bay Area and discovered that the place was poorly maintained. The prices of real estate had increased a bit in the last 6 years, and it looked like it might be an OK time to sell. But there was work to be done. I took a week of vacation, loaded my truck with my lawnmower and whatever tools I might need and drove to Napa. I spent the week fixing as many problems as I could. First on my list was to fix the leaks in the sprinkler system and try to revive the lawn. I hired a painter to paint all of the interior walls. There was a small problem with him. I think he was an alcoholic. He wouldn't eat any of the lunch I provided, but made repeated trips to his truck and by the end of the day he began spilling paint on the carpet. My job was now to follow him around and clean up his spills which became progressively more frequent. I finally had enough, paid him off and sent him home – or to the nearest liquor store. Cleaning up the master bath required a major effort. The walls were coated with a brown film that may have been the residue caused by using it as the marijuana smoking room. A coat of paint covered much of the problem. I had the carpets cleaned but the brown stains from the dog pee would not come up. Many boxes of carpet freshener were needed to mask the odor. The grass had begun to grow again and looked good after being mowed. I hired a neighbor boy to water and mow the lawn every week and thought that area was now covered. My last act before leaving town was to stop by my realtor who was "managing" the rental and finalize the listing contract. It looked like we were going to net about $50K. Renting out one's home can be a stressful undertaking, especially if you are bothered by having someone damage your property through neglect or willful destruction. Actually, most damage can be repaired or painted over, and if one can deal with the anxiety, a lot of money can be made from rentals.

The system definition phase for the SICBM ended and the Air Force asked for bids on the inertial guidance part of the system. The Air Force had chosen the AIRS for the inertial guidance so that part of the contest was over. Autonetics was thought to have a lock on winning this contract since they had a lot of experience with Minuteman and were far along with the MX Peacekeeper system which also used the AIRS. The Air Force was not happy with the possibility that only one contractor would bid and pressured Boeing to enter the completion. Having worked at both places and considering the depth of experience that Autonetics had, I didn't think Boeing had as good a chance as the proverbial snowball. But Boeing is a team player and set about staffing the effort and preparing to put together a credible effort. I was drafted to be the guidance expert. The tracking accuracy for missile velocity during test flights was expected to be very good and could be used in estimating several inertial measurement unit parameters. There was an engineer on the project who had a covariance analysis Kalman filter computer program for estimating many of the gyro and accelerometer parameters. We modified it to use observed (or measured) tracking velocity and produced a graphic that showed how these parameters became observable as tracking accuracy improved. This was an important input to our proposal as it showed that we had some expertise in this area.

The Air Force hosted a contractor conference at Holloman AFB near Alamogordo, NM. The purpose was to pass on to us a number of recent developments in missile guidance and promote some interaction among the various players. I attended along with the proposal engineering manager, Jack, and Jim, one of his management staff. A number of my former colleagues from Autonetics were there so it turned out to be a reunion of sorts for me. There was a "chuck wagon" picnic one evening that was held out on the white sands. I noticed an older gentleman standing by himself and recognized him to be Dr. Charles Stark Draper. He was perhaps the most famous of the early inertial navigation developers and a pioneer in the field. I went over to talk to him. Jack and Jim soon followed and we had a lively and interesting conversation about the early days of inertial navigation. The Draper Laboratories that began as part of MIT where he was a professor are named after him. It was a thrill and great honor to be able to talk with this famous man. We returned to Boeing and began putting the finishing touches on the proposal. There were a lot of different areas to cover. One was IMU recovery after an EMP (electro-magnetic pulse) event. Our recommendation (and the recognized solution) was to use plated wire memory to store critical data that would be needed to bring the systems back on line. The write-up on this was weak and I was asked to revise it. I was new to this technology but tried to write something that made sense so we would look like we knew what we were doing.

During this proposal activity, I began to reflect on my career progress. I had been with the same group in the Aerospace Division since joining the company and was now in my sixth year. I had been given a couple of projects to manage but was not in management. My boss seemed to be content to put me on any task that needed help and had never talked to me about any long term career development. The Boeing News comes out once a week and contains a feature called Employee Requested Transfers (ERT). This is essentially a help wanted section with ads placed by organizations through-out Boeing. I was reading through the job listings and found one from a Boeing Computer Services (BCS) organization. It seemed like a good fit for me and I filed a response. I got a call a few days later and was asked to come in for an interview. I had a good interview with Ben, the group manager and found out that the organization was a staff operation (much like the organization I was in) that sought to place members on projects throughout the company in the Seattle area. If I joined, I would be promoted to management as a pay code 72 (one of the higher line management pay codes, which seemed to be keyed to salary). This would entitle me to a private office and give me an executive parking pass. A parking pass is a prized and much sought after perk for Boeing employees. Most facilities are fenced and one has to pass through a guard gate to enter. Engineers have to park in lots outside the fence and sometimes walk a considerable distance to enter the facility.

Engineers don't have offices and are usually located in "bull-pen" areas. There are also "general parking" passes that allow one to drive in through the gate and park in a lot, but the executive pass gives one access to reserved spaces right next to the building. The new organization also seemed to feel that my recently completed MBA was an asset. The boss of my current organization seemed to have a negative view of the MBA degree; in fact, he had advised one of the people on my GPS project to not get an MBA. In general, managers around Boeing didn't seem to value an MBA much. Few Boeing managers had any formal management training and perhaps felt threatened by anyone who actually had such training. Some Boeing managers were certainly good at their jobs, but relied on OJT (on the job training) for their knowledge of how to manage a project, which tended to be less than comprehensive. I had some reservations regarding the new organization but decided to give it a try. After all, I didn't seem to be going anywhere with my current role so what did I have to lose? The next step was to make the transfer happen. This proved to be difficult. Fred, my current manager, refused to allow the move saying, "We need a cooling off period." He also told my boss-to-be that he and I should meet - and talk. The manager of the proposal I was working on said, "I don't care what group Julian is in as long as he stays on my project." Well, we never had the meeting or the talk, and after a few weeks, the transfer went through. There may have been a limit on how long a transfer could be delayed without taking some action.

Boeing Computer Services

The organization I had just joined was called the Professional Services Group and was a sub-division of Boeing Computer

Services (or BCS). It contained a number of groups with "consulting" or "management" in their names. They provided services to organizations within Boeing but also seemed to have considerable business with other companies. There were about 200 or 300 members of PSG in total. My immediate group numbered about 20 people and was entitled Technology Services – not all that creative a name. Most of the group members worked off-site at various locations around the company so there was little daily contact among them. To provide some interaction, the group sponsored a monthly breakfast to which all were invited. It is interesting that in the beginning, the menu was sausage, bacon, scrambled eggs and fried potatoes. Later this was supplemented and then mostly replaced with a variety of fruits and whole grain muffins, reflecting a growing emphasis on healthy eating at Boeing and in society in general.

As part of my transfer, my pay code changed from "4" (engineer) to "7" (management). I also got an office and an executive parking pass. I continued to work for Jack on the SICBM guidance proposal. Nothing much else had changed except I was now authorized to drive in through the gate and park right in front of the building I worked in. That was nice. We finished and submitted our proposal, then sat back and waited. I took a week off and drove down to California to fix up my house in Napa, as mentioned earlier. When I got back, I found that my boss from the recent proposal effort, Jack, had been frantically trying to get ahold of me. Boeing had lost the SICBM guidance contract competition to Autonetics, but had been awarded a "concept definition study" for the SICBM basing approach, and he wanted me to be his guidance expert.

The SICBM basing plan was to deploy between 250 and 500 single RV missiles on flatbed trucks known as HML's (hard mobile launchers) in the southwest of the country. These HML's would roam the deployment area and move every time a Soviet surveillance satellite passed overhead. The missiles were covered with an inverted "V" shelter and if under attack, would be dropped to ground level so that they might survive an overpressure wave. The HML's would periodically visit a surveyed site for guidance system recalibration and for daily position updates. In the interim, the missile would keep track of its location through land navigation performed by the on-board missile guidance system so that any potential launch point would be accurately known.

I was appointed to be the supervisor of the guidance group and had 3 to 4 engineers reporting to me. The project was short on office space so I was put into an office with Tex, the supervisor of the "systems analysis" group. He wasn't happy to see me move in, apparently reluctant to give up his private office. I had known him on the CLS program, but had little contact although I had once made a wise comment regarding his cowboy outfit that he sometimes wore to work, and that had seemed to piss him off. Anyway, I set about defining the tasks that I had to perform and assigning the work to my staff. My engineers were on loan from the Guidance group, my former home. My former manager had vowed that I wouldn't work in Boeing Aerospace again, but now I'm hiring his people. One of them was a senior person who had been hired 3 or so years ago with one of the highest salaries in the group. He now seemed to be having some performance issues. A supervisor in my former group (not Fred) begged me to try to find a place for him on SICBM. In our initial interview, he talked at length about how we should be using the stellar-inertial system instead of AIRS. The decision to use AIRS had been made so this was not relevant, but I thought it might be a useful exercise to get his views on paper so I assigned him the task of writing a memo making the case for stellar-inertial. When I checked with him a few days later, he had produced nothing but proceeded to give me another lecture on the virtues of stellar-inertial. I had to send him back to his home organization. In the years since he had been hired, he never got a raise and was eventually let go.

One of the tasks that the project was on the hook to perform was to determine how many of the surveyed sites known as PBM's (pre-surveyed benchmarks) would be needed. This was apparently a difficult problem to solve, and Jack, the project manager, had even asked in an all-hands meeting for volunteers to suggest solutions. I ignored the issue since I considered the job to belong to Tex's systems analysis group and knew they were zealously guarding this responsibility. However, my immediate superior, Jim, asked me to look at the problem. I had a reputation for solving tough problems and decided I couldn't turn down the assignment. It was important to limit the number of PBM's (at least initially) since each PBM took a while to set up, requiring star sightings and accurate measurement of gravity, and there were only 2 crews who were currently trained to do this.

My wife and children had gone to California to visit her parents so I had a whole weekend to myself to work the problem. A key element in analyzing the problem was to determine how much (i.e., what percent) of the time an HML would be at a PBM. This time included a periodic recalibration and short daily visits for a position update. In coming up with this number, I even included time for the truck to slow down and stop and then speed up after leaving. A conservative estimate of the time turned out to be about 10 percent. This meant that for a force of 500 HML's, an absolute minimum of 50 PBM's would therefore be required (although this number could be reduced by lowering the 10% requirement). I settled on a slightly larger number of PBM's (i.e., more than 50) as the starting point minimum in order to avoid waiting times and any traffic congestion. The overall objective was to have enough PBM's so that the probability that a missile would be at a PBM was fairly low since the attack would likely target all of the PBM's first and use whatever remained to cover as much of the deployment area as possible. I worked most of the weekend on the problem and in the end, produced a number of results that showed missile survival probabilities as a function of a number of parameters including number of PBM's, attack size, percentage of the time an HML is at a PBM and total basing area. I wrote a memo with my results and distributed it to my immediate boss, the program manager and various interested individuals.

It took a few days for the shit to hit the fan. Recall that one of the Boeing taboos is that one should not or could not interfere in someone else's work. I received a visit from one of Tex's engineers. He had red-marked his copy of my memo and opened the discussion with a heated attack on some of the details of my report. It was soon clear that he didn't understand my analysis (if he had even bothered to read it). I tried to explain what I had done, but he clearly wasn't listening. His position seemed to be that he was the expert and that he needed to discredit my analysis (even though he had not been able to come up with anything). Whether it was right or wrong didn't seem to matter. The fact that I had been asked to look at the problem had also somehow been lost in the mix. He became increasingly emotional and was clearly not interested in what I had to say. I could see that we were getting nowhere and asked him to leave. Tex, who was in the office during the discussion, later reported to Jim, our boss, that I had abused his employee. I decided against discussing the issue with Tex since it was he who had sent his engineer to attack me and seemed to be invested in making me look bad. I instead wrote him a note that simply said that I was dismayed that he was unable to understand my clear detailed analysis of the problem, even to the extent of being confused about some basic definitions.

I didn't hear anything more on the subject until Jim called me into his office. He told me that he had sat down with Tex, and they had gone over my memo "line by line." A few numbers had been updated but my analysis had been approved and accepted. Jim had read my memo, understood it and agreed with my solution. The PBM problem had been solved.

There had recently been a reorganization in my home group. My immediate supervisor, Ben was being promoted to second level management. His former boss took a position as a "senior scientist," reporting to Ben. Two new groups had been created. I was selected to be the supervisor of one of these, which I named the Applied Physics Group. I was given 3 or 4 people to manage as a start. My mission included recruiting and bringing in more members to build up the group. I now effectively had 2 jobs, one supporting the SICBM Program and the other managing a technology support group. I was very busy but found the situation stimulating.

My BCS boss had figured out how to bend the rules to build a large organization of technical experts and advisors that could be deployed to provide support to projects throughout the company. He persuaded candidates to join the group by offering them management pay code (PC7) status, an initial pay increase and, of course, a prized parking pass. Many of those who accepted this offer and joined the group were not known to be superior performers, but after being with the organization for a while they began to shine. It is amazing what giving people a little respect and status can do.

When referring to engineers and managers in this narrative, I use the male gender form for pronouns. This is not because of any sexist point of view on my part but because nearly all of the engineers and managers that I worked with over the years were male. The number of female engineers that I encountered could be counted on the fingers of one hand. Apparently, there is something about engineering that is unappealing to women and few pursue it as a course of study and a career. There were quite a few women, however, who chose to be computer programmers and many found success in this discipline.

Land navigation accuracy was an important part of system performance. I had earlier developed a guidance system computer simulation that included error models for the major error sources and modified it to run in land navigation mode. I hired an engineer, Tim, into my new group. His job was to run the program and study the performance for a number of different guidance systems for various navigation scenarios. He seemed to like the assignment and enthusiastically jumped into it. He made lots of computer runs and the printouts began to accumulate in large stacks in his office. At one point, he stated that this was the best job he had ever had. As the SICBM concept definition study effort began winding down, I met with Tim and informed him that it was time to start putting together a report on his study. He had a puzzled look on his face as I told him this. It seemed he might need more detail so I outlined the contents and scope of the report. It should have been obvious that it would cover a description and objectives of the task, how he had proceeded, detailed information on the results that were obtained and a summary covering whatever conclusions were reached and insights that might have been gained. A couple of days later, Tim came to my office and handed me his report. It was a few penciled sentences on a half sheet of paper – no figures or tables of data. I thanked him and said I'd get back to him. I needed some time to get over the shock. Tim had worked at the company for some 20 years and didn't know how to write a report or document his work? Incredible! If one completes a project and doesn't write a report so others can see what was done and perhaps use the results, it would have been a complete waste of time and money.

My organization had a training budget, and everyone was entitled to travel once a year to somewhere in the country and to attend a class or seminar. Tim was scheduled to attend a course in Portland the next week. I pulled his airplane tickets, cancelled his course and told him he needed to find another job somewhere else in the company – if he could. He was gone within the week – his former organization took him back. I admit that I came down hard on Tim. Normally,

I would have worked with him to structure the report, identify the needed pieces and even write most of the narrative, but his first attempt to produce a document was so inept and insulting that I just wanted him gone. My punishment was to spend the next 3 weekends in the office going over what he had done and writing the report myself. In the end, I produced a 40 page report that covered the work done and addressed the major issues associated with land navigation. I reflected on my role in creating this mess and concluded that it was mostly because I had failed to check Tim out well enough before bringing him on board. The lesson to take away from this is that there are a significant number of degreed employees in the company who cannot or will not produce the documents and reports required of them.

An important part of my role as a BCS supervisor was to make sure that everyone, including me, in my unit had work and was "billing." Boeing might win the SICBM basing contract, and I would probably be offered a job on the project; however, it could be weeks before the decision was made and I needed an assignment in the interim.

There was a project for the US Army, called AOA that was just starting up in the Aerospace Division (my previous home). AOA stood for Airborne Optical Adjunct. What is that? Well, an optical sensor that measures angles to a target was to be installed on a Boeing 767 aircraft. This aircraft would fly at high altitudes to minimize atmospheric refraction, and the sensor would search for objects, such as ballistic missile reentry vehicles and decoys that emit radiation in the infra-red spectrum. An infra-red sensor has a number of advantages over radar as it can track hundreds of objects at the same time and is a passive sensor, emitting no energy of its own. The main disadvantage is that it does not measure range. The project task that I was interested in was the design and development of the algorithms and flight software that would process the sensor data and establish tracks for the observed objects.

The manager of the organization developing the flight software had come from a BCS group that provided technical support to projects around the company. This group contained a lot of PhD's who looked for advanced research assignments that would allow them to write technical papers. They thought of themselves as an elite group. The story was that this manager had difficulty managing people to the degree that he was told to move on. This was supposed to be a "last chance" assignment. I was somewhat acquainted with him in that I had seen him in a number of meetings over the years. I contacted him and offered my services. He immediately put me to work and asked me to check on the progress that a PhD from his old organization was making in designing a Kalman filter for tracking the objects that were acquired by the sensor. I initially thought that giving me this assignment showed a lot of respect for my ability and expertise, but more about that later. After chatting with the analyst, I concluded that what he was doing was unlikely to work. He had decided to implement the filter in polar coordinates – a very unusual and more complicated approach. I suspected that he was doing so hoping to come up with something different so he could write a paper. Furthermore, he didn't seem to be making much progress. I reported back to my new boss with my findings and advised that a traditional xyz coordinate system implementation would be a lot easier to create and have much less risk. The analyst was gone in a few days, and

THE BOEING COMPANY - SEATTLE ➤

I was assigned the task of coming up with the equations and procedures for my recommended approach. Also, my new boss asked me to check on the work of another analyst, Tom, who was doing something very similar to my task. A few days later, the boss called us both into his office and had Tom pitch his work with me offering criticism. Next he had me pitch my stuff and Tom critiquing it. The boss's management style was now much clearer. Apparently, he liked to assign the same task to 2 different people and then have them debate over how to do it. The project staff soon gave him the nickname of "Dr. Zinger," a reflection on his management style and a rhyme with his last name.

The Guidance Group (my former organization) and the Systems Engineering Group in Boeing Aerospace had the responsibility of providing technical advice and support to the project. Dr. Z liked to micro-manage and asked a lot of questions about what they were doing and how they were going to proceed. When he insisted on knowing the details, he received a rather arrogant reply and was told "don't worry about it." This pissed-off Dr. Z, and he began looking for ways to cut them out of the project.

There were 4 main areas where technical challenges needed to be addressed. They were:

(1) The infra-red angle sensor scanned the field of view every ten seconds. All of the points of light that were detected were to be saved and stored after removing any sightings of the fixed stars as contained in a star catalog.

(2) The saved sightings were to be grouped into sets of 5 that

145 ➤

were judged likely to be consecutive sightings of the same object.

(3) These "candidate tracks" were then to be processed in a trajectory forming algorithm to determine if they had the characteristics of a ballistic trajectory. For the candidates judged to be incoming objects, initial position and velocity state vectors would be computed.

(4) The tracks, defined by a position and velocity vector, were to be put into a Kalman tracking filter and updated with every additional sighting made by the infra-red tracker.

The first task was perhaps the easiest, requiring only an efficient data storage and retrieval method and ability to manage the star catalog. The second was going to take some ingenuity to identify probable tracks and select a manageable number. We were required to handle threats containing as many as 2500 objects so the number of track candidates assembled from 5 sightings each could be extremely large. The third task was likely to stress the throughput capacity of our flight computers which were early-eighties technology. The Guidance Group people had come up with a way to create trajectories from the sightings data which they called the "f and g series." While this seemed to work, it was extremely slow and could not meet the through-put requirements of the flight computers. For the fourth task, Dr. Z decided to go with Tom's somewhat unusual version of the tracking filter, which may have come from outside the project and seemed to be the one favored by the customer, the US Army.

After a few weeks on the project, I was appointed supervisor of the Algorithms Group with 7 or 8 engineers reporting to me. They all worked in a secure computer room and would be developing many of the needed algorithms, a simulation of the system and eventually the flight computer code. There were some personnel issues going on with 3 of my new reports. They were on loan to the project and were unhappy with the way they were being treated by their parent organization. I'm not sure how it came up, but transferring to my organization seemed to be the solution. The switch was made. They were promoted to the management pay code (PC7) and received parking passes so they could drive in through the gate and park in the lot next to the building where they worked. They now seemed quite happy.

I began looking at the third task, i.e., forming tracks from the sightings data. Pierre-Simon LaPlace, a brilliant French mathematician in the eighteenth century, had come up with a method to compute an orbit from three or more azimuth-elevation angle sightings. I applied this method as described in "Fundamentals of Astrodynamics" written by 3 professors at the US Air Force Academy. I used the curve fitting method that I developed on the Safeguard Program to compute the angle and angular rate data I needed and programmed the LaPlace method as found in my reference. I told Dr. Z that we should start developing the computer code for our simulation, but he said it was too early. I disagreed and decided to proceed on my own. Back at my BCS office, I had a computer terminal that was connected to the AOA computer so I started stopping off there after work every day for an hour or so and began creating a simulation of the third task. I had a number of pieces from previous projects that would be needed including the aforementioned curve fit program, a ballistic missile simulation, an airplane simulation and some Kalman filter code. My approach was to initialize the ballistic trajectory with the estimate from the Laplace method and then refine it with the Kalman filter. The most accurate initial estimate occurred at the 3rd of the five sightings I was using, so that was the starting point for my initial position and velocity state vector. I flew the trajectory backwards in time to the first sighting, did my Kalman filter update and then flew forward making a filter update at each of the remaining 4 angle sightings. I used the method of computing partial derivatives that I learned during my early days on Minuteman to propagate the state vector covariance matrix forward in time. This all seemed to work well and I incorporated all of these elements into a simulation that took 5 sightings from the AOA aircraft sensor, formed a trajectory and refined it with the Kalman filter. I later added a reentry phase with my version of a Kalman tracking filter to track the missiles all the way to impact. I was confident that my approach was making optimal use of the sightings data and was producing the most accurate trajectory initialization possible. Its use of computer resources was economical, much faster and more accurate than the "f and g series" competition, as we would eventually see.

My BCS boss, Ben, was aggressively trying to grow our organization and pressured me and the other supervisors to recruit new people. I found the best way to find qualified people was to get referrals from those already in the organization. They only recommended individuals they had worked beside and were known to be productive and superior performers. We did run ads in the Boeing News, but this didn't work very well. Applicants usually were underqualified and didn't have the skills and background asked for in the ad. I interviewed one fellow who listed on his resume just about every skill and experience possible in the computing world. Having been burned before, I asked him if he was good at writing reports and documenting his work. Also, would he mind showing me a few examples of his efforts to document his work? He never came up with anything but did stop by one day and told me I had to hire him, and if I didn't, he was going to my boss and have him force me to hire him. That was the last time I saw or heard from him. Apparently his meeting with my boss didn't work out as he expected. My group began to grow and eventually numbered 15 at the peak. My job was to make sure everyone was employed and "billing." A billing rate over 90% is very good and is computed from total available hours minus time for vacations, training and sick leave. My unit (including me) always billed over 90% and always had the highest rate in the Technology Services group. I was able to place a large number of people including those from other parts of our BCS group on AOA and at one time had 12 or more on the project.

Dr. Z came from BCS and recognized the advantage of hiring my people even though they cost a bit more (because of some double overhead). If someone didn't seem to be working out, Dr. Z simply requested I find someone else, and I would have them gone within the week and a new person on-board in a few days. If the project hired someone, they were stuck with that person whether he worked out or not. Getting rid of a non-performer could take a lot of time and effort.

One of my people was working as the main analyst on the candidate selection task and was reporting great progress and excellent results. I never found out what actually happened, but I suspect he was faking his results. Anyway, when he was exposed, he was ordered off the program immediately. This was very rare behavior. Most engineers are scrupulously honest in reporting their progress. However, later events exposed other character flaws in this individual.

Security was tightened as the project progressed. Initially, one could access the project mainframe computer over a telephone line, but this soon changed as many aspects of the project became classified "secret." Project offices and workspace occupied most of one floor of a building and access was controlled by "punch locks." There were a number of smaller work areas and the room that housed our main computer that were accessed from the main room. Access to all of these was also controlled through punch locks. Our security rules required lock combinations were to be memorized and were not to be written down. It was a serious violation of the rules if one were to do so. For most people, this did not present a problem and they dutifully follow the rules. One exception stands out. An analysts on the project, who happened to be in Ben's BCS organization (though not in my group), went on vacation. Another worker on the project found it necessary to go to his desk to retrieve some information and stumbled upon a list of all the lock combinations. The access codes were immediately changed and when the employee returned from vacation, he could not enter the project work area. He was notified through his home organization that he had been "fired" from the project. I thought he might be fired from the company, but that did not happen. He had to learn the hard way and was lucky to still have a job. Everyone else got the message that they needed to follow the project security rules.

When Dr. Z finally gave us the go ahead to start writing the code, I was already finished with my part and delivered it to the simulation builders. My input to them was a computer program that contained models for the flight of the AOA aircraft, an incoming ballistic missile trajectory and my solution to forming trajectories from groups of 5 azimuth and elevation sightings from the optical sensor. This gave them a major head start in getting something up and running.

After a few months, most of the required features had been added to the simulation including the reentry phase tracking filter. Tom, who had developed the filter, had left the program. Getting this part of the program up and running then fell to a young engineer who had recently transferred into my BCS group. He had no Kalman filter experience and was having trouble making this part of the program work. Dr. Z asked me to help him, so we spent an afternoon going over the code and running a few test cases. After a few adjustments to the state vector process noise and other initialization parameters, the filter began accurately tracking RV's using simulated angle data from the IR sensor.

As we made progress, two of the new managers in my group who were building the simulation seized the initiative and presented a plan to convert the FORTRAN code in which it was written to the Pascal computer language which was the one required by the flight computer compiler. This may seem like an obvious step, but it was not in the plan. Clearly, this would make the process of transferring the code to the flight computers much easier. I found it interesting that by elevating engineers to the management pay code and giving them a few perks, they took responsibility for their tasks, made suggestions on how to improve the process and seemed to care about doing a good job. In time, four managers in my group would be promoted to being supervisors on the program. Of course, making them supervisors was easy to do since they were already in the management pay code. The Aerospace Division was very reluctant to create any new managers and there was also a great reluctance to put anyone who was not already in the management pay code in a management job. I was happy to let my people help them out.

The Guidance Group had been cut out of the algorithm development process but refused to give up. They began claiming that their "f and g series" method of trajectory initialization was more accurate that my Kalman filter approach and challenged me to a shootout. I thought it might be a fun exercise so I accepted. I was confident that nothing could beat the Kalman filter approach. A few false starts seemed to be needed to ensure that both simulations were doing the same thing such as modeling a rotating earth and using an elliptical gravity model. In the end, it was very close; however, my Kalman approach produced a more accurate trajectory estimate by a few feet. That ended the controversy and we never had to get around to dealing with the fact that their approach was too slow.

One of our tasks was to develop navigation algorithms that would compute transformations between LL (locally level) coordinates and ECI (earth centered inertial) coordinates and other related coordinate systems. An engineer was brought on board for this early in the program although the activity would be one of the last areas to be addressed. He seemed to be relaxed and enjoying his long tenure on the program. As the time came for him to begin providing his inputs, he had nothing to show for the time he had charged to the task. The most likely reason seemed to be that he didn't have the knowledge or skills (and maybe the desire) needed to do the work. When this became clear, he was booted off the program. He was replaced with another engineer who seemed qualified and began developing the navigation equations. Lesson learned: Don't bring someone onboard until you actually need them and when you do, immediately require some work product that can be inspected to determine competence. It fell to me, however, to review all of this work. The navigation equations included some quaternions which were 4 dimensional vectors defining the ECI to LL coordinate transformation and would be used in referencing the star catalog. The quaternions would also be used by the IR sensor supplier, Hughes Aircraft, to determine sensor pointing directions. This was all new to me, but I can read technical manuals and quickly became an "expert." I didn't like the quaternion notation being used with its subscripts and superscripts used to denote the "from coordinates" and "to coordinates." The order seemed wrong so I switched the "from" to the "to" in the notation. When the guidance people found out about this, they went ballistic and mounted a nasty attack on me and my work. To me, it was a trivial notation issue and I would have gladly changed it back if I had been asked in a respectful way. They, however, were still angry over being cut out of their oversight role on the program and chose to make this a major issue. I decided to play along just to see how far it would go. This all culminated in another shootout held on the Guidance Group's home court. I presented a tutorial on quaternions and showed the relationship between a quaternion and its conjugate. Most of the audience didn't seem to know much about the subject and were mostly just curious. The other side offered a stack of memos written by former members of the group that used the notation convention that they preferred. However, I had seen other reports that favored the notation that I was using. At no time did they state what their objections actually were, i.e., I think they just wanted their notation and not mine. I began to realize that I had the final say in this and let my version stand. I later asked the Hughes IR sensor representatives what their preferences were. Their response was that they were not sure and would use whatever we gave them and see if it worked. If not, they would try to use the quaternion conjugate. In the end, I wasn't sure whether the guidance people understood that we were only talking about the notation or if they thought that I had actually changed the math, which I had not.

The guidance people were not done with me. I wrote a memo that defined a number of coordinate transformation matrices relating the various coordinate systems we were working with, such as: earth centered inertial, earth centered earth fixed, locally level, sensor, etc. I received a memo in reply from one of the guidance engineers stating that all of my transformations were wrong. This area was my long suit. After many years of doing this, nobody did it better than me. I rechecked all of my calculations and confirmed that they were correct. I wrote the engineer informing him of this. I never heard back.

The AOA algorithm and flight computer code development was moving along on schedule. All of the issues were being effectively dealt with. Boeing Aerospace program management was very happy with our progress and recognized and appreciated the key role that my managers and engineers were playing. We were given several awards which consisted of a plaque and photographs taken with upper management. There were some reservations in Boeing Aerospace regarding these awards. They didn't like the fact that they had to rely on BCS people to make things happen and didn't like our PC7 status. On the other hand, they were happy that their program was a success.

I went to a seminar in San Francisco that was also attended by the vice president of Missile Systems, the division of BA that ran the AOA Program. I went over and introduced myself (he probably knew who I was). He invited me to go with him to get lunch. We had a pleasant enough lunch except that I felt obligated to pitch my organization to him. He did not respond. Whoops! Well, they liked us, but they didn't like us.

My BCS boss, Ben, was an outgoing and outspoken person who sometimes said things in meetings with other groups that got him in trouble. He also had one quirk that seemed really odd. When he was meeting with me and/or others in his office, he would sometimes thrust his index finger up his nose and furiously turn it back and forth. We would, of course, be trying to maintain a blank look on our faces. I figured it must be a compulsive thing – he just couldn't stop himself. He developed a habit of frequently not coming in to work, claiming he was "working at home" although we had no clue as to what kind of work that would be. This was strange since most managers were known for coming in early and leaving late. I was never a fanatic about this but made it a point to come in earlier than most and not leave until there were only a few people still there. It seemed like a good idea to set a positive example.

The company awarded annual raises to all employees. The percentage raise for the management paycode was usually a little more than for the other employees, but there were "holdbacks" and "reserves" so it sounded like there was more money than was actually available. My boss and all of us supervisors would meet in a joint session to decide on how to distribute the available funds. This was a flawed process. If someone was to get an above average raise, someone else would have to get a below average raise. Our employees were, of course, by definition all "above average." Each supervisor would describe the performance of each of his people and tended to embellish what he/she had actually done to make sure that his employees got a good raise. This was necessary because the other supervisor felt they needed to do the same thing. It was also important to get letters of recognition and awards for your people during the year so that there would be tangible evidence of good performance. Another consideration was whether an employee was "underpaid." There was a sense that this should also be a factor in determining the person's raise. A much fairer system would be to let each supervisor have what his employees had generated and work with the boss to allocate it within the group. I came to believe that we should not try to "fix" an underpaid employee's salary because this would unfairly take away raise money earned by the other employees. Besides, the underpaid situation was the results of years of past performance and decisions made by the employee and his/her previous management.

I was asked by Ben to try to find a job on AOA for Matt, one of his other supervisors. I arranged an interview for Matt with Dr. Z, and he was hired and put in charge of the requirements group. I thought that I had done a good thing by helping bring this about and was surprised to find that a few weeks later; Matt was standing in for Dr. Z in project review meetings. I was the most senior and most knowledgeable supervisor on the program, and it seemed that this role should have been mine. This just didn't smell right. It took me some time to figure out what had happened. Boeing had a program called EXPO which stands for executive potential. Its purpose was to identify individuals who might move into upper management and provide them with appropriate training and also allow them to interact with various corporate executives. Matt was in this program. Matt and Ben were tight, and I was sure that Ben and Dr. Z had conspired to provide Matt with this exposure as part of his EXPO training while cutting me out of picture. I was collateral damage. They probably thought that I would never figure out what had happened.

This incident caused me to reflect upon my situation. I was less than happy with other aspects of Dr. Z's management style. He did all of the interviewing, hiring and firing for the entire group. I was one of his supervisors, but I didn't get to do much supervising. The doctor also felt free to reassign my people whenever he felt like it without consulting me first. He was also pressuring me to come up with a way to address "aerosols and chaff," which were a part of the threat, although they seemed to be of minor importance. There was one secret document that mentioned them but supplied no detail. There didn't appear to be a way to get any additional information and there was not enough available to do a decent analysis. I suggested he ask the Army program office for some additional information. This did not happen, but he kept demanding that I come up with something. I decided that it was time for me to move on and sent Dr. Z a letter of resignation from my supervisor role. I did offer to continue to provide technical support as needed. And so, my participation on the AOA program began to come to an end.

Our organization was officially known as Technology Services and was part of a larger organization called BCS Professional Services Group. The other sub-groups or units served many non-Boeing customers and provided consulting and management services to these clients. We, on the other hand, worked almost exclusively with Boeing Aerospace and provided engineering and technology expertise. Someone decided that there should be a reorganization and that we should be cut loose and become part of BASD, the Boeing Aerospace Support District, also a BCS division. This made sense on paper and the organization name was a good fit, but we were a much different organization than the one we were joining. BASD provided mostly tech workers and computer programmers to the Aerospace Division. They didn't do much analytical or engineering work. After the transfer and after they had a chance to get a good look at us, BASD top management was amazed that nearly everyone was in the management pay code. This didn't set right with them, and their gut reaction was to charge HR, the human resources organization, to pursue a massive demotion of our personnel. As this played out and some time went by, HR began to realize that there actually were reasons for our status and it made sense at some level. But they were not deterred, and we were tasked to write several memos justifying our status. They were faced with a problem in that our people had received a promotion, and many were, in fact, working as line managers. A wholesale demotion of personnel without cause might be hard to explain and would certainly cause problems for our customers. We were never given any rational as to why the demotions were necessary or made any sense at all. My guess is that there was a feeling that the privileged status of our people caused envy and was felt to be "unfair." Also, part of HR's mission was to limit the number of people in the management pay code because they typically did not bill and were an overhead expense. There seemed to be a basic ignorance in HR as to who we were and how we operated. Our PC7's all billed and brought in a lot of revenue to BCS, in fact, we had often been referred to as the "cash cow" when we were part of PSG. Our PC7's were not an expense but a major revenue producer.

At first, it looked like HR was going to come in with their meat axes swinging and cut us down, but the process suddenly slowed. Our PC7's became aware of what was going on and wrote a memo to the BCS VP and some of the upper level managers. One of the more pointed paragraphs was:

> "Job promotion is one way that BCS tells you that you are appreciated and a valued employee. A demotion makes the opposite statement. The total professional staff of 40 employees is being demoted from PC7 to PC6. There doesn't seem to be a communication problem in BCS; management's message is coming through loud and clear."

It was signed by 9 of our PC7's. After reading this, the BCS VP sent out a memo stating that he had authorized a 3 month study of the situation. I don't think any further action was ever taken although the issue never died. Two and a half years later, memos were still being written asking for a set of requirements that one would have to meet to keep PC7 status. The bottom line was that HR had effectively shut us down. We could no longer offer anyone PC7 status to join our group. Our key recruiting tool had been taken away from us, and we stopped growing. The process of natural attrition would probably reduce our numbers over time.

There were still people who wanted to transfer into our group even without getting the previously available benefits. They transferred straight across and remained at whatever pay code they had been. I had 2 pay code 4 engineers in my group and had a problem at raise time since both had performed well. I couldn't give one an above average raise without giving the other a below average raise. One of them, an expert in Ada and Unix, had been working on a proposal effort and had been something of a hero. He had been put in charge of a team and wrote a key part of the winning proposal. He expected an above average raise. When I told him the bad news that he had gotten an average raise, he gave me a menacing look. I told him the reason, but I don't think he believed me. Managers worry about unstable employees and I remember having an eerie feeling. He left the company soon after that, but called a couple of times and asked to use me as reference, which was fine with me since he had been a good worker. One night as I was watching the local news, I saw him on the television wearing an orange jumpsuit. He had been arrested for murdering his wife and dumping her body in the mountains. A few years later, I saw a television show called "Forensic Files" that featured the story and provided details of the crime. The couple had both wanted a divorce, but he thought it was going to be "too expensive." He solved the problem by brutally beating her to death. He had tried to clean up the bedroom crime scene but botched the effort and the investigators nailed him with that evidence.

There were actually very few personnel problems with the people in my group. The engineer who was faking his results and was bounced from the AOA program managed to get a sexual harassment claimed filed against him. We were able to smooth this over with an apology letter from him and a sincere promise to never do it again. My advice for success in the workplace is: (1) become an honest person, (2) show a positive attitude, (3) learn how to do something useful and become good at it and (4) don't try to play or con your boss. He probably saw through you the first time you tried it. Did I mention: Don't do anything really stupid? Perhaps that goes without saying.

Every two or three years we would receive the news that there was a budget crunch and we needed to reduce expenses. The first thing to go was the training budget. It was wise to take one's training early in the year and avoid losing out. Several times we were asked to prepare a lay-off list. This was always puzzling to us since our people were almost always working on a project and bringing in revenue. We historically were able to bill out over 90% of our time and deserved our reputation of being a "cash cow." Some individuals were periodically let go in different parts of BCS, but these seemed to be people who were not contributing or able to perform useful work. The lay-off process started with a "warn" notice meaning that the person had 30 days to find a job somewhere in the company. I had encountered one of these individuals on the IUS program some 8 or so years earlier. He had a PhD in mathematics and should have, by this time, established a solid career doing something of value for the company. He came by looking for a job. Ben asked me to pick him up and find him some work. After all, he had a PhD. He was probably in this predicament because he had never taken his job seriously and applied himself. I remembered his antics and odd behavior when he was on IUS and didn't want to deal with any of it. No one else stepped in to save him and he was soon gone.

We stayed with BASD for a little over a year and were then transferred to a group called the Scientific Computing and Analysis Division. This is the organization mentioned earlier that Dr. Z had come from and the one that had a lot of PhD's who were focused on doing original research work. This was, I suppose, a better fit for us since it was a more professional organization. With every reorganization, one can expect there to be a physical relocation. Boeing had a very active facilities department and a wide choice of locations. We were moved to a new building in the Factoria Shopping Mall in Bellevue and fairly close to our new parent organization. This building was 4 stories high with underground parking. My office was on the top floor and faced west giving me an unrestricted view of Seattle and the Olympic Mountains. That was nice.

Ben began to exhibit some erratic behavior. He may have felt some stress from the PC7 debacle and the reorganizations. He may have been having trouble in his personal life. Anyway, he held a meeting once a week with his supervisors to go over new information and resolve any business issues. The meeting was scheduled for an hour. He began spending the entire time scolding us and did not address any of the business issues. I began a subtle campaign to have all of us supervisors meet without Ben and deal with the issues that needed to be resolved. Ben stumbled upon our meeting once and demanded, "What's going on here?" We had to make up something on the fly. The company had recently adopted an evaluation process where each employee filled out a form and rated his manager. The results of Ben's evaluation were mixed. Several people had slammed him hard on his management style. He was furious and demanded to know who had criticized him. The ballots were anonymous, and no one knew who had written what nor was anyone observing his reaction going to admit to making any negative statements.

The head of our new organization was a respected scientist and seemed to be a capable manager. In order to get to know his new employees, he invited each of us in for a brief interview. During my interview, we got around to talking about Ben. I decided to give him the full story. He must have gotten the same information from several others. Shortly thereafter, the organization head began to put pressure on Ben to give up his management role. Ben had offended a lot of people over the years and had not built any alliances. No one came forward to defend him. He was eventually forced to step down. Our organization was split up and the groups were attached to different parts of the Scientific Computing and Analysis Division. My group, Applied Physics, became part of Applied Mathematics. Oh well, what's in a name?

Every year since I joined BCS, all management employees were required to fill out a form called a Management Progress Planning and Evaluation (MPP&E). Initially, this was a fairly trivial and innocuous process. The employee typically mentioned things like serving the customer's needs and acquiring some additional training. The forms were duly filled out, filed and only looked at briefly as part of one's year-end review. Eventually, there was some pressure, that must have come from somewhere up the line, to improve the process. I agreed that this was worth doing and came to believe that there was potential that was being overlooked. I decided to revise the form and added several areas of performance that defined the role of our personnel in some detail. These included things like suggesting improvements to customer's processes, taking responsibility for meeting schedules and managing one's tasks, etc. I identified 6 areas where our people could step up and help the customer succeed. I think this helped our people understand what their role and mission was. It also included the usual part about increasing one's skills and knowledge. I sent out a memo (with form attached) to everyone in my group with instructions to fill it out and send it back to me. I got about a 50% compliance with this. After you have been a manager for a while, nothing fazes you. Your job is to make what is supposed to happen, happen. So I called each of the non-respondents and asked them to fill out their forms and send them to me ASAP. This worked for all but one guy. I scheduled an appointment for him in my office and when he arrived, I got on my computer. We discussed and filled out each entry, after which I had him sign it. A manager does whatever it takes to make it happen.

This effort was well received by my boss and the other supervisors in the organization, who either copied what I had done or did something similar. It seemed that we had made real progress. However, this was not going to last. It may have been that our efforts got noticed or it may have simply been the normal evolution of events. The next year some major changes were ordered. After years of being ignored, someone in upper management took an interest in the MPP&E. Instead of being an individual bottom-up exercise, it became a decidedly top-down approach. The VP of our division of BCS filled out his MPP&E and identified the areas he was going to emphasize, the activities to be addressed and something about expected results. His direct reports were to pick up on this and structure their own MPP&E's to identify activities to support the VP's plan. This was to flow down to the next level and eventually down to the lowest rung of the management ladder in the organization. In essence, everything that we were to do would be in support of the VP's plan. I preferred what we had before.

I had noticed over the years that information had a hard time moving up the line. A line manager was usually well aware of what his/her people were working on. The second level manager had some knowledge of what those below were doing but had to depend on the line manager's diligence and cooperation in passing information along. A good second level manager would require frequent and detailed information about what was going on. Some did; many didn't. Beyond that, there wasn't much detailed information that moved up to the 3rd level and above.

Upper management felt the need to have accomplishments to point to at the end of the year. One way to fill this requirement was to bring in the latest training program or "magic bullet" that would in theory inspire the workforce. The list of these programs is quite long and includes CQI, TQC, WCC, SPC and the 5 S's. The long form names for the first 4 of these are constant quality improvement, total quality control, world class competitiveness and statistical process control. The 5 S's was a Japanese inspired program that dealt with keeping a clean work area with the S's standing for things such as sorting and sweeping. Every employee was required to take a day long course and managers attended training for a week. There was some good information in the training and some ideas that could be used to improve performance. Unfortunately, there was no implementation plan. No support services were made available and there was no follow through to see whether anything was being accomplished. Individuals could take the initiative and try to come up with a plan to do something. Some did, but most didn't.

The BCS EXPO (executive potential) program sought to identify and develop individuals who could become upper level managers. The program kept a low profile and few people outside of upper management even knew about it. I petitioned my boss to nominate me for the program and I was accepted. I had been burned by this program when an EXPO had edged me out of an assignment on the AOA program. There seemed to be some advantages to being in the program, and I didn't want to be missing out on something. I was 54 years old at the time and had no illusions about moving up in the company, but what could it hurt? There were about 40 people in the current group. We went to a lot of classes on leadership, managing, etc. We also engaged in some community support activities such as: "Invent America," which was a program for elementary school children that encouraged them to invent things. The inventions were judged in regional and eventually national competitions. There were also a lot of gatherings where company leaders were invited to speak. The one I remember best was when a young, dynamic vice president was the speaker. He talked about the importance of working with people you like and getting along well with them. He said, "If I find myself working with someone I can't get along with, I get rid of them." This statement was quite shocking, and I thought that that was a luxury most of us couldn't afford but appreciated his candor. It seemed to be working for him. There were also breakfasts with the company vice presidents. Twice a month, we went to the headquarters building at Boeing Field and had breakfast with one of the company vice presidents, after which there was a question and answer period. The CEO was even the guest at one of these. This was a great opportunity to meet the leaders in the company. I was favorably impressed by all of them (well, there was one who for some reason would not take my questions). I always had a couple of questions ready for them. I had learned in business school that the purpose of one's question was not to demonstrate how smart you are but to give the speaker an opportunity to expand on a topic of interest to him/her.

The BASD group employed an individual whose job was to establish and maintain contacts with all Boeing Aerospace projects and look for opportunities for BCS to support them. He described his role as that of a "trapper." I got to know the trapper quite well over the years. He began coming up with assignments for me where project managers needed an independent view of what was happening on their own projects or an outside opinion on what should be done. This appealed to me since I always thought it would be interesting to be a detective. The manager of the "virtual systems" project asked me to take a look at what his people were doing and to do some research to see if there were any similar projects being pursued by other companies. After meeting with the project manager, I got the impression that he didn't trust his employ-ees and wasn't really sure what they were doing. I began by talking to some of the people on the project to find out what was going on. I was assured by the project personnel that what they were doing was unique and there was nothing else out there like it. Every time I thought I understood an aspect of the effort I got the same response, "No, it's not like that." I be-gan to understand why the manager was concerned. A large part of the problem here seemed to be that the workers on the project didn't respect their manager and thought they could just give him some vague answers and then do whatever they wanted. They apparently hadn't thought about the possibility that the manager could bring in help from outside the project.

The purpose of the "virtual system" (as I understood it) was to serve as an interface between modules containing code written in the Ada programming language and the computer and operating system so that Ada modules would be portable to different computing systems. I enlisted the help of some researchers in my parent organization, and we found that there were several companies doing something similar. We produced a report discussing the many activities in this area in some detail. A critical finding was that the BCS "virtual system" did not reference or specifically align with any of the standards being developed to promote Ada portability. My customer seemed very pleased with the report, and I'm sure he now knew a lot more about the environment his project was working in and what should be done. My hope was that he now knew enough about what was going on to effectively take control of his project.

The trapper located another possible source of business on

a project that was trying to get some government research money to look at ways to identify Soviet mobile missiles. These missiles were to be placed on trucks or railroad cars. I had become interested in neural networks and thought that this technology might have an application on this task. The idea was that a drone would fly over an area where missiles were thought to be located and use neural network based technology and pattern recognition to identify the missiles. One of the PhDs in my organization had successfully built a neural network based method for identifying aircraft parts in the warehouse. I could not, however, convince the project manager that neural network pattern recognition could be very effective in solving the problem and that this would be of interest to the customer. The PhD did generate some interest at DARPA and the Navy for the work he was doing and we traveled to Washington, DC to talk to these agencies. They were very interested in what had been done so far, and it looked like we should follow through here. There was one problem and that was that the PhD didn't have a secret clearance which was essential for these types of projects. I pressured him to submit his application, but he never would. He eventually left the company. I concluded that he had something in his background that he wanted to keep covered up and that might cause his application to be denied. Could it also be something that he had hidden from the company?

The AOA project that I had worked on for many years turned out to be a great success. The flight computers being used were early 80's technology and suffered from a lack of storage capacity and speed. It was necessary to divide the flight software into 3 parts, each on a different computer. This was a difficult and challenging task but the team, led by people from my group, took it on and made it work. The customer, the US Army, was thrilled with the airborne system that could acquire ballistic objects at long range and track them to impact using data from an angle measuring infra-red sensor. The system met or exceeded all expectations.

After less than 2 years with the Scientific Computing and Analysis Division, it was time to reorganize and move again. I don't know what the rational for the move was this time, but it seems we didn't fit in well anywhere. We were something of an orphan group. We were again attached to BASD and reporting to our former boss who had now moved up to be a fourth level manager and was now the head of BASD. Boeing had many different locations on the eastside, many of which were rented during business expansions. Some of them had acquired colorful nicknames. For a time, we were in the "milk of magnesia" building which had blue glass sides and windows that produced an eerie blue glow in our work space. We had also spent time in one of the 2 "flash cubes," which, yes, resembled a camera flash cube. This time we were relocated to "stonehenge," one of 2 new buildings on a cul-de-sac with a park at the end featuring several stone pillars reminiscent of the ancient monument. Word got out that our new leader was going to reconstitute our group and reinstall Ben as our boss. Apparently he knew little or nothing of our former leader's antics. I was not surprised since he had not previously been much involved in our activities and never visited our work area when we had been part of his organization in the past. When I heard of this, I immediately gave the big boss a phone call and told him I didn't want to work for Ben again and gave him my reasons. My friend, the trapper, was equally aghast at the possibility of having to work with Ben again and also made a call to register his complaint. I don't know if he got any other calls, but Ben did not come back. Case closed.

In the Stonehenge building, the offices for supervisors were small, being no more than 7 feet wide. All of the saved space seemed to have been collected in the spacious complex occupied by the division manager. There was a roomy reception area and a large office for him. There was also an adjoining large area that wasn't yet being used for anything. I suggested it was being reserved for a bowling alley. Our new second level boss had been in the organization for some time, and we had all had some contact with him. He appeared to be a rising star. I was surprised to find that he was in awe of us and thought we represented some kind of an elite group. Our reputation had preceded us. Normally, we would have let the "new guy" work things out in his own way. However, I could see that he was struggling and tried to help him wherever I could. Unfortunately, he remained anxious and could never seem to get comfortable. He left work one day and checked himself in to a hospital. When he eventually returned to work, it was somewhere else.

One of the other supervisors in the group asked me if I would help a manager who was trying to develop a new business initiative. The manager was the one who had been the radar guy on the CLS project that I worked on when I first came to Boeing. We had not parted as friends, and he apparently had some reservations about contacting me directly. I felt bad about the situation and realized I had not given him a chance. I had relied on what one of his former coworkers had said about him. Although this information had turned out to be essentially correct, it should not have been the basis of our dealings. Anyway, I decided to help him out. He wanted to use the computer simulation I had developed on AOA that could acquire and track ballistic missiles using angle only data. I turned the program over to an analyst in our group who read in the code using an optical character reader and downloaded it to one of our Apple scientific desktop computers. He got it running in short order and generated results for a variety of incoming missile trajectories. I was invited to attend the meeting where the results were presented to an audience of military and industry people. I was surprised to find that Fred, my former boss when I was in the Guidance and Navigation Group, was leading the effort. He completely ignored me and made no effort to talk to me. Apparently, he was still angry that I had left his group although he was OK with using my work.

Management Control Systems Audit Group

After I completed the EXPO program, I was offered an assignment as an auditor in the Management Control Systems Audit Group (MCSAG), a group that was part of the Boeing Corporate Division. Most people, when they hear the word audit, think of accounting and ledgers, but these audits were quite different. They were what I would call process audits. They looked at how things were being done and identified problems. It sounded like an interesting change of pace. It was a one year assignment so I took a leave from my current position and joined the audit group.

A few other new auditors joined at the same time, and we started by attending a week long course on what auditing was and how it worked. After the training, each new auditor was paired with an experienced auditor for that first assignment. I was teamed with a woman who had been on a number of audits, and we were assigned to look at systems engineering on the F-22 fighter program. Unfortunately, after a couple of weeks, she began having some medical issues and left the program. The F-22 audit was put on hold for the time being. The audit group manager thought an audit of "technical compliance" would be a more appropriate place for me to start. They were shorthanded when it came to experienced auditors so I was assigned to be the lead auditor even though it was really my first audit. Another new auditor, who was perhaps 10 years my junior, was assigned to help me.

I began the audit by learning about technical compliance and found that it is an essential function that is used on proposals where it identifies standards and requirements that apply and develops a plan on how these areas will be addressed on the contract. The audit had been requested by the technical compliance (TC) organization so we had their full support. We began by meeting with our TC contact and then branched out. We talked to a large number of TC support workers and some proposal managers. After a while, a clear picture of the issues and problems began to emerge. Some proposal managers said they didn't know what the TC person was doing or even was supposed to do. They complained that they were forced to have this person on the proposal team, but he was just "burning budget." There was no formal training for TC people so the learning was OJT. The more senior and experienced TC personnel seemed to be good at the job, but even then there were gaps in knowing what to do. I discovered early on that the best way to find out what was going on was to interview both sides – the service supplier and the customer. Between the two stories it was fairly easy to get a clear picture of what was really happening. After we had done our research and learned about as much as we could about TC, it was time to write our report and make our recommendations. I asked my audit partner to independently write up his conclusions. I would do likewise and we could then compare our views. My jaw dropped when I read his summary. He thought the problem was that the TC organization didn't treat its people well enough. "Wow! OK, I guess I'm on my own here." It seemed clear to me that the major problem was the lack of formal training for the TC personnel. A training course needed to be developed. This course would identify the responsibilities of a TC person when supporting a proposal and a checklist of what to do and how to do it. It should be made clear that before beginning to provide support on a proposal effort, the TC person should sit down with the proposal manager and go over what TC was supposed to do and what documentation would be produced.

There was another area under the TC umbrella that we investigated. When a product is delivered, all applicable documentation including requirements, test results and production details need to be available for the customer to inspect before signing-off and accepting the product. The B-2 program had a good and effective process for identifying and recording all of the applicable and required items as they were created during the production cycle. These documents were then easily accessed and made available during the delivery process. Another program had a hit and miss technique resulting in every product delivery being a "crisis" event. Our recommendation to this program was to adopt a process patterned after the B-2 method.

I wrote a report with my findings, conclusions and recommendations and submitted it to the TC organization. I also made a presentation of my results to a couple of VP's from TC. One was an older distinguished looking gentleman who was 75 years old but, as we had heard, liked his job so much that he decided not to retire. The other VP was quite a bit younger and an ordinary looking fellow. Everything seemed to be going well until I mentioned something about problems with "level 3 drawings." The younger VP jumped up from his chair and headed toward me while shouting, "I fixed that problem!" I thought I was dead. The older VP rose from his chair and intercepted him while saying, "Now, now Dan, you know there are still some issues there." Except for this near disaster, my report was well received at TC. My main contact there was extravagant in his praise and said that my findings were just what he thought. So chalk up a success here. I may have had some doubts as I was writing up my findings – this being my first attempt at auditing – but I seemed to have nailed it.

For my next audit, I was assigned to look at systems engineering on the F-22 fighter program. Lockheed was the prime contractor but it was a joint project and Boeing had a large piece of the action. This is the audit I had started on when I first joined the audit group, but it had been postponed due to the illness of the lead auditor. I was now the lead auditor and was assigned a different assistant than I had on the TC audit. I had seen systems engineering being performed on many projects over the past few years and welcomed this opportunity to really dig into the discipline. We began by reading the SOW (statement of work). The SOW included a WBS (work breakdown structure). I noticed that every task in the WBS required that a "deliverable" be produced. A deliverable is documentation showing results and details that provided evidence that the task had been successfully completed and is the end product of the task. What a great idea! This showed that the thinking on how to ensure performance had evolved and that both customer and contractor had made progress in how to make sure the work was getting done.

The F-22 project had a lot of aspects that were classified "top secret." I didn't think that I would need to know any TS information, but it was requested that I apply for a clearance anyway since I would be in the vicinity of where this information was being used. This resulted in a visit from a couple of government investigators. I'm sure they had "profiled" me and had decided on any areas where I might be vulnerable and might be exploited to disclose classified information. They were mostly interested in my alcohol consumption, but after a long and detailed interrogation, they seemed satisfied that I didn't have a drinking problem.

Before we were really able to get started, it was announced that a team of systems engineers from the aerospace and military airplane divisions were going to perform an audit of F-22 systems engineering. It didn't seem to matter to them that we were already doing an audit, if they even knew. They fielded a team of older veteran engineers who appeared to have a lot of wind in their sails so it seemed to me to be a waste of effort to try to deter them. I did the best thing I could think of and asked if I could join their team. They seemed to be "cock sure" they knew what they were doing although I'm sure they knew virtually nothing about auditing. They never asked me for any help. Because I was part of the team, I was able to attend all of their meetings and keep track of what they were up to. They spent a few weeks on-site and then scheduled a meeting with the F-22 program manager to present their findings. I attended the meeting and was anxious to see how their conclusions would be received. They had prepared a 4 to 5 page handout filled with bullets and sub categories. The F-22 program manager found problems right away and a lively argument ensued. We never got more than half-way through the second page before everyone realized that it was futile to go on. The objections to what they had presented were that their conclusions were not well-supported by the evidence that was collected and their recommendations lacked detail and focus. I hung around afterwards and spoke with the program manager. I told him that I would like to go ahead and proceed with my audit and that I was sure I could help him. He said, "Go ahead."

We were back in business. We began by doing basic research to become familiar with the program and how it was organized. The program was built around IPT's (integrated product teams) with each team assigned to accomplish a major program task. A systems engineer was assigned to each team to instruct and advise team members on how to do various tasks and to make sure the team had access to the documentation needed to accomplish these tasks. We started our audit interviews with the manager of systems engineering and branched out to talk with the team leaders and as many team members as we could. My favorite question for each team member was, "What is your deliverable?" It was important for the person to know what he was trying to produce. There were several engineers that didn't know and I assigned them homework to find out. We used the technique of interviewing both the service provider and the user of the service extensively and were able to get a good feel for what the problems were. It soon became clear that the program was suffering from the same problem that we had found with the TC organization. The systems engineers had not been sufficiently trained and did not have a comprehensive view of what support they were to give the team. Also, they had not effectively communicated to the other team members what services and assistance they were there to provide. Consequently, the team members did not know how to do many of the tasks required of them nor did they know where to go for help even though the help they needed was supposed to be right there on the team.

There were a couple of less serious, though significant problems. One was that there was no effective process in place for bringing a problem to someone's attention and getting it solved. Another problem involved a remote Avionics laboratory that was part of systems engineering but required a special clearance to get into the lab. Systems Engineering management had trouble managing the lab because of the clearance issue.

We put together our report identifying the problems and our recommendations for fixing them and presented our results to the program manager. He seemed very pleased. After an audit, you move on and don't necessarily know whether anything was done and if things worked out. An auditor has a lot of power and the audited group is required to implement the auditor's recommendations. That's fine in theory, but it is unlikely to happen unless the audited group agrees with the audit findings. A few months later, I was talking with my neighbor who was also a Boeing engineer. He told me he had recently been reassigned to F-22 Systems Engineering.

His new job was to implement my audit recommendations. They had accepted our findings and recommendations and were going ahead with the changes. I should have followed up on this after a few months to see what had actually been accomplished but didn't.

The audit group employed a number of "subject matter experts" who were supposed to have detailed knowledge of company procedures and policies, as well as, expertise in various disciplines. The idea was that they were a resource that auditors could call upon to help them quickly get up to speed in the area they were preparing to audit, and also be a source of continuing support if needed. The "experts" had nice offices while the auditors who they were supposed to serve were housed in a bull pen area. The message that they were working for us may have had some trouble getting through. On my first audit where I was the junior auditor on the team, we had spent some time with one of the experts who was supposed to know about systems engineering. He addressed all of his comments to the lead auditor and totally ignored me. I thought that this was quite rude. This expert seemed to be arrogant and feel that he was a cut above a mere auditor. When I later became in charge of the systems engineering audit, I didn't use him, besides I probably knew more about the subject than he did. He didn't see that one coming. The manager of the audit group called me into his office one day to talk about how it was going and at the end of our session, asked me if I had any concerns or critical observations. I said that, yes, there was one and that I didn't think the experts (at least the one I named) had the right attitude, that being, that they were there to serve the auditor's needs. At some point, the expert's jobs would seem to depend on whether they were providing a useful service to the auditors. I mentioned that the fact that the experts had nice offices and the auditors were in the bullpen might be a factor contributing to their perception that they were more important than the auditors. I'm sure the manager talked to the expert afterward and asked him about the situation. So, who was he going to believe? Well, I had given him something to think about.

My next audit was "manufacturing process control" in the Fabrication Division of the Boeing Commercial Airplane Group. We started out as a 3 person audit team. My team mates were both younger than me and in their early forties. I'll call them Jeff and Ann. Jeff was near his one year anniversary in the audit group and would be leaving before we finished the audit. Ann and I got off to a strange start. She made a lot of humorous, somewhat nasty comments directed at me and seemed to be enjoying her cleverness. She may have been intimidated by me for some reason that I couldn't identify, or she may have just been trying to impress Jeff. I didn't seem to have any choice so I responded in kind. This verbal "grab ass" went on for some time. Jeff told me he was concerned about the relationship between Ann and me and didn't see how we could finish the audit after he left. I told him not to worry as the nonsense would stop as soon as he left and Ann didn't have anyone to play to. As soon as Jeff left, things changed immediately, and Ann and I quickly formed a business-like and professional relationship.

The problem we were to look at was quality control of manufacturing processes for aircraft parts. A considerable number of the parts that were produced didn't pass inspection and were either sent back for rework or were scrapped. Over time, this problem was costing the company a lot of money. Each of the processes for producing parts was defined by a set of detailed written instructions. Quality Assurance personnel were on the floor observing the manufacturing to ensure that all steps were being performed correctly. Once a month they marked a checklist that indicated whether they had observed that each step in the process was being performed correctly. These checklists typically showed that all steps in the process were being followed, i.e., there was 100% compliance. We discovered that even though some violations of the process steps were being observed by the QA people, a checkmark was usually awarded as a "courtesy" after a warning had been made. Based on these "perfect" scores, one of the VP's of the Fabrication Division had even gone on the company news and stated that there were many processes in the division that had "zero defects." However, his statement was clearly inconsistent with the fact that the levels of scrap and rework remained quite high.

I found a magazine article that described the phenomenon of "rain dancing" which is defined as doing things that look good and feel good but don't produce any results. The QA practice of awarding a perfect score even though the processes produced a lot of defective parts seemed to fit the definition. The term became our favorite word for describing the QA activity although we never used it around the people we were auditing. The term also became popular in the rest of the audit group who seemed to appreciate the humor.

We finished our audit and pointed out that the QA approach being used was ineffective and misleading. We recommended that better training of workers and periodic monitoring of individuals for compliance with manufacturing procedures would be the best way to reduce production errors. This was my last complete audit. I did start on an audit of the handling of hazardous materials in the company. This would have been very interesting, but my time was up and I returned to the job I had before becoming an auditor.

We now had a new boss who came from the computer programming side of the division. He didn't seem to think we were anything special. When my performance review came around shortly after I returned, we had an argument about my accomplishments. He seemed bent on diminishing the significance of what I had done and what members of my group had contributed. I thought he was probably trying to set me up for a below average raise.

BCS - Software Engineering

It seemed like a good time to go elsewhere. I was an EXPO and had served a year in the audit group. This was supposed to put me in a position to be offered a good job; however, no one was beating my door down. I had been offered an unspecified job by the manager of the Technical Compliance group I had audited, but TC was not that exciting to me and didn't seem to offer any interesting challenges. Jack, the program manager I had worked for on the SICBM programs, was now the director of the Software Engineering group in BCS, and I decided to check with him. As it turns out, he was very upset with me for leaving the SICBM project although no one had ever talked to me about staying on the project while I was still there. Apparently, all was forgiven or at least enough so for me to be hired. I was offered a job although it was unclear exactly what I would be doing.

BCS had just purchased a new software development system called P+ from a Canadian company. The biggest problem with software is that it is hard to maintain. Proper documentation is often lacking and going back to find out how it works and to make changes is time consuming and costly. This new method provided for extensive documentation during the development process and because of that, maintenance should be a lot easier. My new job would be to develop metrics for projects using the P+ method and to track progress.

I had no sooner settled into my new office when I got a call from Human Resources (HR) asking me to agree to a pay code change. They wanted to change my line management pay code to either a tech code 6 or a lesser management code 7. Apparently, if I agreed to the change, it was a done deal, but I sensed that if I didn't, it was going to be a lot harder for them. My position was that I would be assigned a line management role in the near future and that if I gave up my current status, it would be nearly impossible to get it back. The HR person argued long and hard, but I refused to budge. I think HR enjoyed screwing-over the employees. Maybe, it was their way of welcoming me to the group. I think the decision was eventually left up to my new boss, Jack, and he said to leave me alone.

Shortly after arriving at my new location, the audit group sent over a nicely framed certificate of appreciation and a presentation ceremony was scheduled. I had my picture taken with the president of BCS and also with 2 directors (my new boss and the one from BASD, my previous organization), the head of the audit group and my new immediate supervisor. This BCS president was a legend in the company. On several occasions when a division was found to have serious performance problems, he was sent in to straighten them out, and he had always quickly got them back on track. He had a gruff-looking appearance, which may have caused people to think he was no one to be messed with and to do whatever he asked. The BCS presidency was probably a reward for his good service and, of course, proven executive ability.

The airplane division had started building a Central Storage Facility (CSF) 4 years earlier. The CSF was designed to be an automated, inventory storage and order fulfillment system for aircraft parts. A contract for the development of the software had been awarded to an outside company. A little over a year later disagreements over scope and feasibility put the project on hold. P+ was brought in as a way to save the project. Our VP asked Jack to write a "white paper" assessing whether P+ had been successful in solving the problem. I was asked to do the research and write the paper. Having just come off my audit assignment, this was right up my alley. I interviewed all the players, made my analysis and wrote the paper, which was then sent up the line. The conclusion was that P+ had rescued the software part of the project but was less than effective in addressing the hardware/software interfaces, but then it was not designed to handle that task. A few people had yet to tinker with what I had written and put their scent on it before sending it up the line. Their changes made it a little harder to read and understand, but the basic message remained intact.

My supervisor put me on a committee to help with a system development methodology being proposed by one of the groups in our organization. The committee was made up of individuals from all of the disciplines in the organization including facilities, accounting and human resources. I sat through the first few sessions to get a good picture of what was being proposed. It soon became clear that the objective was to reinvent the systems engineering methodology but with a BCS slant. This seemed totally unnecessary to me since there was a well-defined process being used in other parts of the company and it would have made sense to simply adopt or adapt that. Most of the other people on the committee had no background in systems engineering and had no clue as to what was going on. I took it upon myself to educate the committee members to some degree and made copies of some systems engineering manuals I had. I passed them out at our next meeting. After this session, the meetings were suspended for a while. Before they were resumed, I had a meeting with my supervisor and told him what I thought was going on. The project was cancelled. My input may not have been the only reason for the cancellation, but I think it may have started a reevaluation of the project.

There were a number of people in BCS middle management who spent a lot time (and for some perhaps most of their time) in meetings. It was hard to see what benefit this produced since the work of the organization was being done by others, and there never seemed to be any meeting by-products that made their way into or affected the workplace. My direct boss was one of the "meetings crowd" and seemed to much prefer this activity as opposed to managing his direct reports. "Three hundred and sixty degree" reviews where the employees rate the boss had just become popular. My boss had about 10 people in his group who rated him in the survey, and all except me had dinged him for not spending enough time supervising them. I personally liked my boss and didn't need much supervising, so his management style was fine with me. In my review, I wrote that he seemed to be trying to do two separate jobs and it was hard for him to find time to do both. I didn't see my comments as being all that positive, but my supervisor asked me to schedule a meeting with his boss and discuss my point of view. I did, of course. Maybe this helped get him some relief from the pressure.

P+ was now the required software development methodology and every project began using it. It was, however, a complex system that required extensive documentation at every step of the way. The decision was made to train a group of "coaches" and then have a coach available for every project to assist in developing these documents. Jack asked me to apply for the coach training. I was definitely not interested because I thought I was way overqualified; however, Jack insisted and I submitted my application. The multipage application form was a long checklist of programming languages and computing systems and asked for the number of years spent on each. There was nothing about experience working with people or any analytical ability which might come in handy in coaching others. I was relieved when I was turned down although I had to wonder how someone as well qualified as me was rejected. This apprentice program was being run out of the airplane division and moved ahead with a first group of 15 or so students. Several "consultants" from the Canadian company were hired to be on site and manage the training of the Boeing people. The first group finished their training. A second group was then selected and began their training.

Someone decided that it was time to reorganize, and that the training program, which had been in BCAG, the commercial airplane group, would fit better in the Software Engineering group of BCS. That was us. Jack picked me to be the manager of the group. This made a lot of sense since I was already a manager and had been one of Jack's supervisors on the SICBM program. Also Jack and I had a long track record where I had come through for him on some difficult problems in the past. He obviously thought that I was a known quantity, and he could trust me to do the job. However, as soon as the announcement was made, there was a steady stream of people meeting with Jack in his office to plead their cases that they would be a better choice than me. One of these was Jim, who was not in management but was to all intents and purposes the manager of the coach training program. He was the guy who had turned down my application to be a coach. He was now working for me. I doubt if any of my competitors knew much of anything about my background or realized that I was probably much more qualified than they were.

I began doing some checking starting with his previous supervisor. I found that Jim was pretty much impossible to manage and they had more or less let him do what he wanted. He and his assistant, Pam, were clearly in control and did as they pleased. That was not going to work for me. It was clear to me though that getting control of this operation was going to be a tough slog. They were located at a different facility so I began spending some time there attending all of their meetings, and I also scheduled interviews to get to know each of the trainees. Jim was feuding with one of the consultants and threatened to not renew funding for her. I found out that this particular consultant was also the local manager for all of the other consultants and deserved some respect. We became friends and I assured her that her funding would continue. My first act had been to take control of the budget, and I now had an ally among the consultant group. My goal was to learn as much as I could about what was going on and be ready for the third class when and if there was one.

I began my research to determine if there was enough demand for coaches to warrant a third class. The biggest complaint about the coach training program was that many organizations couldn't get one of their people accepted into the program. I scheduled a couple of meetings and invited all of the organizations that might use a coach. I found it interesting that many people will not speak up in a meeting, but if you contact them with a phone call afterward, they have a lot to say. I proceeded to put together a plan for another class that including evidence that it was needed and requested the funding. This had to go through our VP. The director of Software Engineering, Jack, and I met with the VP. He criticized my plan quite harshly and said that I had not made the case that another class was needed and I needed to come up with a lot better story. I thought that I had done a good job and didn't agree with him at all, but one doesn't argue with the boss even if it would have made a difference. In situations like this, my first supervisor at Boeing liked to claim that upper management thinks it has to pee on your work product, i.e., mark it with their scent. Maybe the VP just wanted to show he was doing his job. I made a few changes and resubmitted the request for funding. It was approved this time.

To be honest, I really didn't care if it was approved or not. I was 58 years old, and my wife and I had been talking about retirement for a while. Carole had gone back to work 8 years ago when our daughter got her driver's license and didn't need to be driven around anymore. She hadn't worked at a company for 19 years and had to go back to school to update her secretarial skills. Her first job was at Boeing and didn't pay very much, but I knew how to work the system some and she had a knack for identifying all the ways she could get raises. Besides she was a very capable worker and her bosses were glad to help her increase her salary, which became quite respectable after a few years. She left Boeing after 5 years and took a job as assistant to the Program manager developing an ultra-sound system at Siemens. Our children, David and Cheryl, had both graduated from Western Washington University in the past year, and we were now able to boost contributions to our savings. I was seriously considering retiring at the end of the next year.

The third coach training class started with the 15 students that I had selected. I decided to appoint a training manager to define and manage the training program and work with the consultants. Jim and Pam were now completely cut out of this part of the operation. I still needed them for some of the administrative activities and had not figured out a way to replace them there yet. I went in search of a training manager and at first thought that Bernie, another manager and friend in my group, would be a possibility. He was working on a doctor's degree in some education field and it seemed he would be a good fit. I invited him to come into my office, and after talking about the job, I offered it to him. He jumped up from his chair and left without a word. I was shocked by this reaction but later realized that he was one of those people who had petitioned my boss to take my job away from me and give it to him. In his mind, my job should have been his, and now I wanted him to work for me? What an insult! This turned out to be a big mistake on his part. I later offered the job to a training instructor I knew from the BCS course training group and he accepted. After his assignment as my training manager ended, he was offered a major role in a large new BCAG training program based partly on the experience he gained working on my program. Bernie could have had this prestigious assignment. I guess he screwed up.

The P+ coach training program moved ahead and was expected to take 6 to 8 months. At the halfway point, progress was going at a slower than expected pace. I scheduled a "come to Jesus" meeting and informed the trainees that we were running out of time and that the program funding would soon run out. At that time, the training consultants would be gone and everyone would return to their home organizations – finished or not. If they failed to finish their training, I would look bad and so would they. It might be hard to explain to their bosses why they had failed. This meeting lit a fire under all of them, and they began to make excellent progress. The students who were farthest ahead began to finish, and all finished by the scheduled end of the training.

Our consultants from the Canadian company did a good job in helping the students learn the P+ process. There was one minor glitch however. One of them had made a sexist remark in a meeting that was attended by several women. A complaint was filed so I began an investigation. The accused consultant admitted that he had made some inappropriate comments and promised never to do it again. This seemed to satisfy the complainants, and it appeared that the problem had been resolved. However, a week or so later, he did it again. I think the stipulation had been made that if he repeated the offense, he was history. After a short conference with those involved, I went to his office and informed him that I would be back soon with some moving boxes and help him pack up his things. He said, "Right now?" I said that it was so. We wheeled his things down to the back entrance, and I helped him put them in his car. I shook his hand, wished him "good luck" and watched him drive away.

I began getting reports from my training manager and others that Jim was "poisoning the well" by making statements to people outside of the program that the current class was to going be inferior to the previous 2 groups and was making other subversive comments. It is not easy for a manager to reign in an out of control employee. Normally, pay raises and performance reviews instill some fear in the employees and give the manager a lot of leverage. However, the company had delayed the usual year-end raises and the accompanying performance reviews until the following year so that was not yet available.

Boeing Commercial Airplane Group

We reorganized again and the P+ training program now became part of the commercial airplane organization from whence it came. I had started out working in Boeing Aerospace, transferred to Boeing Computer Services and was now finally part of the Boeing Commercial Airplane Group. I had a whole new set of bosses in a large group of about 130 people. The manager of this group decided that we should rate all of our employees and provided us with a form

containing several performance categories such as: produc-
tivity, competence, communication, teamwork, etc. A rating
system of 1 to 10 was used. Most managers like to err on the
positive side and give each employee better marks than he
or she actually deserved so that nearly everyone is "above
average." In rating Jim and Pam, I decided to be honest and
give them the marks that I felt they had earned. Jim came out
below average, mostly because of his lack of a positive and
supportive attitude and, also, his subversive activities. Pam
got rated average because she supported Jim in his negative
behavior. When all the result were collected and combined,
a relative rating graphic of all 130 people in the organization
was made available to the managers. Jim was dead last and
Pam was next.

The economic recession that we were just coming out of had
resulted in a large number of airplane order cancellations and
production had been scaled back. All of the managers were
asked to meet and come up with the names of employees that
the company could lay-off. I offered up Jim as someone who
could be let go. My program and, therefore, his assignment
were ending soon. He had not talked to me about a job elsewhere and I didn't think he had anything lined up. Normally,
a manager would work with the employee to locate a follow
on assignment, but Jim had not indicated any interest in my
help, and besides I would be unlikely to wish him on anyone.
The company preferred to lay-off young workers with little
experience; however, the company had not done much hiring
in the past few years and there were very few young workers
to be considered. Laying-off workers over 40 was a problem
because of recent legislation at the national level that made
the company jump through some extra hoops to do so.

One day, as I was walking through the work area, I noticed Jim with his head down on his desk and stopped to see if he was OK. He was unresponsive so I had our secretary call Boeing medical. They sent out an ambulance and tried to revive him. He was groggy and having difficulty communicating so they put him on a gurney and took him to the hospital. He had gone to lunch that day and apparently eaten some sea food that had gone bad. The verdict was food poisoning. He was better by the next day and soon returned to work. I never received a "thank you."

Jim received a "warn notice" which meant that he had 30 days to find a job somewhere in the company or be given a "pink slip" or lay-off notice and have only 2 weeks to go. I received a lot of email expressing shock that Jim was being let go and making statements in support of him. They, of course, didn't know the whole story. With 2 days to go, one of the higher level managers in my new group stepped in and stopped Jim's lay-off temporarily. I got a call from a manager who was considering hiring Jim and asked me a lot of questions. "Was he reliable?" "Was he easy to work with?" Was he helpful and agreeable?" Let's see, I had him out the door and someone stepped in to stop it. I no longer had a dog in this fight. I gave evasive answers and didn't make any positive or any negative observations in response to his questions. Apparently, this manager had been told to hire Jim, but felt he should go through the formality of checking with me. I was somewhat relieved to be off the hook. Fired employees have been known to go "postal."

Boeing had found few employees to lay-off and now considered a drastic step. Although I had heard it said that an early

retirement offer would never be considered, the announcement of an early retirement program came at the beginning
April. I had just turned 59 and was planning to retire at the
end of the year anyway. The offer was that older management
employees could retire anytime between now and the end of
June of this year. Those accepting the offer would be credited
with 5 additional years of service, which would increase pen-
sion benefits. I thought, "Can this really be true? What a great
offer." I immediately decided to take the offer but decided to
not tell anyone until the last minute. I had until 2 weeks before the deadline to inform HR. Boeing has always had great
benefits. One that stands out is connected to sick leave. The
company transferred the value of up to one week of unused
sick leave every year into an investment fund for the employee. I hardly ever used any sick leave so I got the full amount
every year. Any additional amount over and above the one
week would be paid to an employee leaving the company at
half of one's hourly rate. The company also paid us for any
unused vacation. Boeing gave me full medical coverage until age 65 and also gave the same coverage to my wife even
though she had already left the company.

Although we had officially transferred to the airplane company, we were still located in the same area and building as
Software Engineering. One of the managers in SE had been a
fourth level BCS manager; in fact, he had been my boss at one
time. He had recently been assigned some low level job and
occupied a small interior office across from my first SE office.
When the news of the early retirement hit, he was immediately seen packing up his office and was gone the next day.
He must have been really pissed about something.

When you know you are soon leaving, it is remarkable how all the stress is gone and you feel totally relaxed. I wanted to wait until the end of June to leave because there was no reason not to, and staying on the payroll another 3 months allowed me to put more money into savings. I was thoroughly enjoying myself and refused to tell anyone if I was retiring, pretending to be still thinking it over. I waited until the last minute to tell HR that I was leaving. I was given a couple of going-away luncheons and a large farewell party in the meeting room at work.

After a month or so into my retirement, I received a phone call from my last SE supervisor asking me to come back in as a contract employee. I had no interest in returning to work but thought I would play along with it just to see what was being offered. Apparently, too many key people had taken the early retirement offer and the ability to get the work done had been affected although the problems were mostly in airplane manufacturing. I also talked with my former supervisor's boss and said I was willing to come back in if I was really needed but advised him to run it by the BCS VP I had worked for when I was setting up the last P+ training class. He had not been all that thrilled with my effort to make the case that another class was needed. I wasn't going to go back to work unless I had the support of everyone in the organization. I never heard back. And so my nearly 35 year career as an aerospace engineer and manager came to an end.

Looking Back

MY PROFESSIONAL CAREER that began in 1960 was now over. I had been hired to be a research engineer by Autonetics fresh out of college with a degree in mathematics. I soon found out that I didn't know how to do anything that was of value to my new employer. I guess they were used to this and were willing to spend a few years educating and training me.

In the next few years, I was given assignments that required me to learn about a broad range of topics related to inertial navigation. When my bosses realized that I could and would get things done, I was often called upon to take on difficult tasks and may have caught a lot of "dirty jobs" that they couldn't get anyone else to do. But these jobs usually had a lot of interesting aspects and I learned a lot from them. I never turned down any job and never really thought about doing so.

The work environment was often challenging with a lot of activity and noise going on all around me. In becoming a productive worker, I learned to concentrate and block out most of the distractions. I learned early on that visiting and just trying to kill time made the day drag on and seem even longer.

Immersing myself in my work made the hours zip by, and it was soon time to go home or pursue some interesting social activity.

Computer programming turned out to be a valuable tool. I created a lot of simulations covering many different guidance and navigation topics over the years. I kept these with me and found the opportunity to use this large and growing toolkit on many occasions and was able to get a running start on several projects.

As my career matured, I found myself able to solve virtually any problem related to missile trajectories and inertial navigation. My intuition had grown over the years and seemed to always direct me in the right direction and toward the solution. I became a firm believer in the principal of "Occam's Razor" which states that the best solution is also probably going to be the simplest and found that often to be the case.

It is ironic that as I became more and more successful in solving engineering problems, I was forced to move into management to attain the career fulfillment that I desired. The outward signs of success such as a window office and reserved parking were not available to engineers no matter how skilled and valuable they may have become. The nice thing about engineering though is that it is a very objective business. If you can get it done, people notice. If you can't, you are soon found out. Management is much more subjective. Success is not always obvious. I can recall cases of managers whose projects were in deep trouble, and then received elaborate praise when they finally succeeded. A better manager would have gotten the job done from the beginning and no one would have thought it was anything special.

Engineering is an exercise in logical thinking and finding solutions to difficult problems. It relies on one's ability and the expertise and understanding gained over the years. A successful career will show a long record of consistent and dependable performance. Management is quite different. Management's job is to identify the work that needs to be done, acquire the needed resources, assign the work to individuals who are capable of performing it and make sure that the work is being done in a timely manner. But perhaps the biggest challenge for a manager is in having to deal with many different types of individuals with a variety of agendas. Many (perhaps most) of those who cross one's path act in a self-serving manner. The interests of the project and the company are a secondary focus for them. Most managers want to move up the corporate ladder, but the number of higher level slots available is limited. For the few who actually make it into management, only one in ten or twenty can ever move up, resulting in a very competitive environment. It is necessary to spend a lot of effort managing one's own image and the perceptions of others and being on-guard as to what moves others are making. Competition and politics are an ever present fact of life.

In the end, I am able to look back at my career as a manager and engineer and the many challenges I faced and enjoy the successes that I achieved. I take satisfaction in the fact that I was able to make many significant contributions and that I was always able to live up to my own high standards.

www.ingramcontent.com/pod-product-compliance
Lightning Source LLC
Chambersburg PA
CBHW071148050326
40689CB00011B/2023